W9-BBR-780

SUCCESSFUL REENGINEERING

SUCCESSFUL
REENGINEERING

DANIEL P. PETROZZO

JOHN C. STEPPER

VAN NOSTRAND REINHOLD
An International Thomson Publishing Company

New York • London • Bonn • Boston • Detroit • Madrid • Melbourne • Mexico City
Paris • Singapore • Tokyo • Toronto • Albany NY • Belmont CA • Cincinnati OH

Library of Congress Catalog Card Number 94–8559
ISBN 0–442–01722–4

I(T)P Van Nostrand Reinhold is an International Thomson Publishing
company. ITP logo is a trademark under license.

Printed in the United States of America

Van Nostrand Reinhold
115 Fifth Avenue
New York, NY 10003

International Thomson Publishing
Berkshire House, 168-173
High Holborn, London WC1V 7AA
England

Thomas Nelson Australia
102 Dodds Street
South Melbourne 3205
Victoria, Australia

Nelson Canada
1120 Birchmount Road
Scarborough, Ontario
M1K 5G4, Canada

International Thomson Publishing
GmbH
Königswinterer Str. 418
53227 Bonn
Germany

International Thomson Publishing
Asia
221 Henderson Road
#05-10 Henderson Building
Singapore 0315

International Thomson Publishing
Japan
Hirakawa-Cho Kyowa Building, 3F
2-2-1 Hirakawa-Cho
Chiyoda-ku, Tokyo 102
Japan

RRDHB 16 15 14 13 12 11 10 9 8 7 6 5 4 3 2 1

Library of Congress Cataloging in Publication Data

Petrozzo, Daniel P.
    Successful reengineering / Daniel P. Petrozzo, John C. Stepper.
        p.   cm.
    Includes bibliographical references and index.
    ISBN 0–442–01722–4
    1. Organizational change.   2. Strategic planning.   3. Industrial
management.   I. Stepper, John C.   II. Title.
HD58.8.P487   1994
658.4'063—dc20   94–8559
                                                               CIP

To Jennifer
For her unconditional understanding

—Daniel P. Petrozzo

To Anne-Marie
For her inspiration and support

—John C. Stepper

Contents

Preface

There is an old story of three blind men describing an elephant. The first man, on grabbing an ear, exclaims, "It is a large, rough thing, wide and broad, like a rug." The next man, on holding the trunk, says, "No, no. It is straight and hollow, like a pipe." The third man, in turn, holds a leg and says, "You are both wrong. It is mighty and firm, like a pillar." Like the three blind men, business leaders have for years been attempting to understand their companies' problems by examining or revamping only pieces of them. As the story goes, "These people will never know an elephant."

This book is designed as a road map, a detailed guide, for the comprehensive change of a business. It is aimed at helping you assess and solve your current business problems by redesigning your processes, organizations, and information systems. In the current business vernacular, such a complete business redesign is referred to as reengineering.

Reengineering is popular because it promises to deliver corporations from the quagmire of inefficiency, high cost, and dissatisfied customers. Unlike the incremental improvements offered by quality management programs, reengineering holds out the hope of working with a clean slate. Such promise, backed by a handful of success stories, has vaulted reengineering into the business mainstream. *Fortune* magazine featured reengineering as the "hot new managing tool" in August 1993, and *Reengineering the Corporation* by Michael Hammer and James Champy was on the *New York Times* best-seller list for months.

Yet such popularity has a price: reengineering has become faddish. A typical sentiment was expressed in the *Fortune* article, "If you want to get something funded around here—anything, even a new chair for your office—call it reengineering." Thus, the term reengineering has become enigmatic, leaving business managers unsure of its definition and even less sure of its real benefits.

Two articles have laid a solid foundation for future work on reengineering. Michael Hammer's "Reengineering Work: Don't Automate, Obliterate" (*Harvard Business Review*, July–August 1990) and Thomas Davenport and James Short's "The New Industrial Engineering" (*Sloan Management Review*, Summer 1990) gave a name to the type of work we had been doing for some time and provided a framework for the body of knowledge available up to that point. After personally implementing large reengineering projects and working with or sharing information with dozens of redesign groups in different companies, we saw the need to offer detailed instructions and recommendations to others engaged in such work.

Despite the promise of reengineering, we recognized that most reengineering projects fail. Hammer and Champy's estimate is that *50 to 70 percent of reengineering projects fail to meet their objectives.* And even the "successful" projects, such as the oft-touted redesign of Ford Motor Company's accounts payable process, can take five years or longer to implement. Common reasons for these failures and delays include:

- Lack of top-management support (a champion) for the project.
- Poor understanding of the organization needed to support the new design.
- Inability to deliver the necessary technology advances.

In diagnosing the inadequacies of major improvement programs, no book that we have read addresses the critical role of technology in creating fundamental change. Popular works by Davenport, Deming, Peters, and Hammer all provide insights on the need for leadership and on the importance of being process-oriented, but they pay scant attention to the role of technology. *Successful Reengineering* will serve as an in-depth guide to reengineering *all three* components of a business: processes, organizations, and systems.

Throughout the analysis, redesign, and implementation sections of this book, we stress two themes: concurrent engineering and evolutionary implementation. Some standard reengineering case studies (e.g., the redesign of Kodak's product development cycle cited in *Reengineering the Corporation*) are, in fact, applications of concurrent engineering principles. A survey paper on concurrent engineering was coauthored by a colleague of ours, Dr. James Pennell, for the Institute for Defense Analyses (IDA), in which he defines concurrent engineering as:

a systematic approach to the integrated, concurrent design of products and their related processes, including manufacture and support. This approach is intended to cause developers, from the outset, to consider all elements of a product life cycle from conception through disposal, including quality, cost, schedule, and user requirements.

We have found that the design and development of a process and supporting systems is best served by using ideas found in many concurrent engineering projects, such as multifunction teams and systems prototyping. As described in this book, the use of concurrent engineering methods in combination with a reengineering mindset is a powerful force in understanding and redesigning a business environment.

In addition, the evolutionary implementation methods we describe apply to the development of the new business design: processes, organizations, and systems. We espouse an iterative approach similar to the Shewhart Cycle (Plan, Do, Check, Act) used in improving processes since the 1920s. Shewhart and the legions of quality improvement experts who have followed him have contended that significant improvements could be made if one followed his four steps:

1. Look at the available data and form a theory or model of how things (a process, for example) work.

2. Based on your understanding, do something you think will improve the situation.

3. Check whether the new data conform to your existing theory or model.

4. Advance your understanding by modifying your model to account for the new data.

We apply these steps to the implementation of a new business design. After developing an understanding of the problems within the current environment, the reengineering team members will implement their new design in successive iterations. They will use tools to model their new design and to help them choose between design alternatives. After each iteration, the actual results will be compared with the expected benefits, and the team can then alter the design accordingly. Having strategists, analysts, and engineers working together from the outset and implementing the new design in increments allows a reengineering project to realize benefits more quickly than with traditional (i.e., functional) methods. In turn, producing results quickly allows the project to validate elements of the new design and make necessary adjustments early in the project, thereby reducing risk.

This approach to reengineering is made possible by a convergence of orientations, attitudes, and technology. Only within the last ten years or so have businesses begun to focus on processes instead of functions (despite the exhortations of Deming and Juran for more than fifty years). Within the last five years, many business leaders have recognized the importance of creating multifunctional, empowered teams to support these processes. And, only now is the technology available to support radical, holistic changes in a business. The methods and tools described in this book, from microworlds and scenario-based training to process simulation and work flow software, represent key advances in the business outlook and technology needed to support both the *reengineering* process and the *reengineered* process.

Successful Reengineering is an implementation guide for reengineering projects, and is intended primarily for these groups of people:

- Senior managers.
- Full-time members of a reengineering team.

- Technology professionals assigned to support reengineering efforts.
- Human resource personnel expected to solve people issues.

However, the book also will help educate and prepare anyone who may be impacted by a reengineering project.

Daniel P. Petrozzo
John C. Stepper

Acknowledgments

Writing a book is one of those intense efforts of creation that requires you to give up part of yourself. Thoughts, personal anecdotes, and attempts at humor are all laid bare on the printed page. Though it is sometimes difficult to have others examine your work, feedback and criticism are necessary parts of the book-writing process. We would like to thank those who helped us.

Many individuals played key roles in reading and commenting on various versions of the manuscript. We would like to thank Mike Salisbury for reviewing nearly every version of all that we wrote. Jennifer Petrozzo provided tremendous assistance with many of the topics related to human resources. Anne-Marie Brillantes' insistence on clarity and style helped make the book more readable. Jim Pennell, our good friend and colleague, provided the inspiration for, as well as critical input to, the material on simulation and design of experiment. Andy Sealock, Bill Huston, and Steve Verderese provided helpful comments on various portions of the manuscript.

We would also like to acknowledge several individuals who played a role in the creation of the book. Bob Argentieri expressed confidence and enthusiasm in our ideas and worked with us during the difficult formative stages of the book. The energy and creativity of our editor, Jeanne Glasser, helped bring the project to fruition. Finally, a special thanks to all the staff at VNR who worked on the many aspects of production, from artwork to copyediting.

1

Introduction

Business problems that were recognized in the 1980s still exist. Despite pitches for quality, empowerment, and "world-class" service and countless publications and consultants promoting these concepts, most programs aimed at making extensive business changes have failed.

A wide array of excuses are given whenever a quality program is tried and fails:

"The top management are not committed."

"The people here just don't care."

"The competition is already too far ahead."

The real reason is that Total Quality Management, quality councils, and process improvement programs simply offer incorrect solutions to many of the problems encountered. You are reading this book because you sense the same thing.

The following two profiles—composite sketches of the many managers we have worked with—illustrate the frustration felt in many companies when quality improvement programs do not yield the anticipated results.

Connie Roberts is the head of an organization responsible for taking customer orders, completing those orders, and ensuring they are properly billed. Connie's organization is also responsible for the day-to-day relationship with all of the company's customers. Over the last several years the quality of the services that Connie's organization provides has deteriorated. She is well aware of this because she must personally respond to the angry customers. At the same time, the organization is growing by leaps and bounds, having doubled in size over the last few years. Connie had to add people to process orders and answer customer complaints, which did not seem to be a problem at first because order volumes were going through the roof. Now the order volumes are tailing off, but even with all its employees the organization cannot keep up with the work.

Connie has a hard time figuring out what the problem is. She knows the computer systems that are used by the people in the organizations are inadequate and never seem to get better. She knows that they are contributing to the quality problems. To offset the burden the computer systems have created, Connie employs a large staff of people responsible for process improvement. These are quality problems, she reasons. They need to be solved through quality techniques. Besides the process improvement people, there is a quality person per functional subgroup, and quality improvement teams abound. This commitment to quality has existed for a few years. The people seem to enjoy the empowerment, but the bottom line is that operational costs have increased, and the customers are still unhappy. Perplexed and disenchanted, Connie rationalizes the situation by believing that she has been fooled again by the latest management techniques and buzzwords.

Peter Daniels is a young computer systems professional working in the organization that develops systems to support Connie's employees. The products that are developed in Peter's group never meet the expectations of the users in Connie's organization. Lately, there is no time even to worry about Connie and the needs of her organization. Development cannot keep up with the demands for new services that the sales and marketing folks want to offer.

These problems persist in spite of the fact that the quality bug has really bitten Peter's organization. The head of the division has sponsored an organization-wide quality council for the last few years, and is even an active member of it. The quality council is considering organizational values, mission statements, and processes. Nothing seems to be working. Peter thinks to himself that there must be a better way.

There are thousands of people like Connie Roberts and Peter Daniels in companies around the world. More than a few of our readers can relate to their situations. Have they been led astray by bad management programs? Not necessarily. The problems that their organizations are experiencing are fundamental to their businesses, and there is absolutely no way that a business will become successful using the quality path exclusively. The message is *not* that quality programs are unimportant; rather, they are only tools that are not sufficient to address a business's deep-rooted problems.

There are many business problems with symptoms like those described above where radical, holistic changes are required. Think about driving in New York City (or any old city). As you drive down the street, you encounter a vicious, unavoidable pothole every block or so. There is continuous construction—an obvious sign of improvement—but what are the chances that the roads will be perfect the next time you visit? Zero, of course. Is this so because the repair people are inadequate, or because repairing roads is so inherently complicated that finding workers with the correct skills is impossible? No, the real problem is that the original designers did not anticipate that the infrastructure would need to support the vast number of vehicles that are constantly using the roads.

Typical business processes, organizations, and supporting systems—the business infrastructure—also were designed for purposes that no longer exist. In many cases the situation is worse than that of the roads of New York City because the business problems have been neglected for several decades. It is no wonder that quality programs do not fix the problems; the problems must be addressed holistically. Each process, organization, and computer system must be analyzed to see if what it is doing maps with what your business must do to survive. There must be a strong emphasis on technology in this analysis. In many cases, technology is actually a root cause of the current problem, but the proper application of technology will enable you to solve the problem.

A DEFINITION OF REENGINEERING

For a holistic view of business problems and a means of obliterating those problems, companies from Aetna to Xerox are turning to reengineering. We define the term as follows:

> Reengineering is the concurrent redesign of processes, organizations, and their supporting information

systems to achieve radical improvement in time, cost, quality, and customers' regard for the company's products and services.

This definition is similar to those offered previously (e.g., in *Reengineering the Corporation*) but stresses two points that are critical to successful implementation: the interplay between processes and the structures (both human and technical) that support them, and the need to redesign *all* aspects of the business concurrently.

This definition scales well; that is, reengineering and the principles of its successful implementation can be applied to companies of all sizes. Whether working in a local retail store or a large multinational organization, managers all over the world can apply the same approach to problem identification and resolution.

HOW THIS BOOK IS ORGANIZED

As the primary purpose of this book is to serve as a reengineering implementation guide, the information is presented in the order in which a project will be carried out. This quasichronological order is referred to as phases. There are four phases, spanning the project from the beginning (the initial idea to attempt a radical improvement) to the end (when the new infrastructure allowing for continuous improvement is in place). The phases, described here by the key activity going on in each of them, are not unlike those of most projects that you are already familiar with:

I. Discover

II. Hunt and Gather

III. Innovate and Build

IV. Reorganize, Retrain, Retool

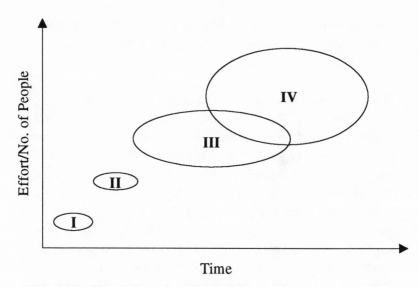

Figure 1-1. The phases of a reengineering project.

The phases are presented in roughly chronological order, but this is not meant as a precise statement of how a reengineering project should be managed. The time spent in each phase is affected by factors such as scope, complexity, and urgency. Also, there is an overlap of some activities across phases. As shown in Figure 1-1, the four phases serve as a useful guide to the approximate time and effort that certain aspects of tasks should take.

Discover

What type of ingenious thoughts serve to motivate holistic change within a business? The cartoon character Wile E. Coyote comes to mind. Mr. Coyote is pretty creative; a little light bulb frequently appears above his head. Ignoring the fate that Mr. Coyote ultimately experiences in his pursuit of the roadrunner, one must admit that he is an "ideas" guy. The pursuit of reengineering usually is inspired by a little light bulb going off in someone's head. However, the actions that one subsequently takes must be carefully evaluated. The fate that usually awaits Mr. Coyote is an undesirable outcome and must be avoided.

The scenario usually goes something like this:

> There is a significant problem. The top management in the division recognizes that something is seriously wrong. Like the prototypical Connie Roberts, they realize that nothing that has been tried seems to work. They usually get some of the "smart" employees to investigate. If management is aware of some of the concepts that support reengineering, they may order these people to go fix the world. It is more likely that these lucky, or unlucky, "smart" souls will venture off to examine an area of concern to management.

The "discover" phase of the book describes the role of leadership and team selection in reengineering projects and provides techniques for identifying areas of a business that are good candidates for reengineering. The chapters on business assessment and setting the project scope will guide you in selecting candidates that offer excellent potential for improvement as well as realistic chances of implementation.

Hunt and Gather

American life in the late 1700s and early 1800s had an interesting aspect: many brave people engaged in expeditions. Just imagine what Lewis and Clark had to experience: every day they were trying to find the lay of the land; challenges awaited them at every turn; there were hostile natives, unfamiliar or unexpected topography, and the constant need to remember their purpose. It is a safe assumption that they probably wanted to turn back a few hundred times. Nevertheless, they persevered. Because they were the right people for the job, they were able to survive, to understand, and to create maps for future travelers.

Analyzing a complicated, broken business operation is not unlike going on an expedition. Workers must study and understand the problem at hand. Like an explorer, the analysis team will constantly be in uncharted territory. Its challenges will include finding friendly people who will help with the exploration, fending off political attacks, and understanding a multitude of complicated, interrelated problems. At the end of this phase, the team will know how to decide on a plan for the enhanced business process and will know which supporting information technology is required.

Innovate and Build

An *innovation* has been defined as "something that deviates from established doctrine or practice." The essence of innovation in a reengineering project is the *rethinking* of business processes, given modern advances in technology and management practices. In addition to avoiding the problems identified during the earlier analysis of the current environment, this phase provides established principles to guide you in creating the new design.

The bulk of this phase is devoted to the methods and the tools needed to implement the new design in an iterative fashion. The methods described include ways of preparing for the new design, developing software, and comparing different design alternatives. The tools range from mathematical tools to technologies. In this phase, the methods and tools are combined in an evolutionary approach that is critical to successful reengineering.

Reorganize, Retrain, Retool

In school we were taught the three R's, which are the basics for all primary educational systems in the United States. If the three R's are not satisfied, then the educational system, the student, or both have failed. This maxim also applies to projects where the goal is to reengineer. Reorganize, Retrain, Retool is a different set of the three R's, but our ability to satisfy them is a barometer of our success. Frequently, this phase of a project is labeled transition or integration. These are fine labels, but the work itself is accomplished by using the three R's. Much like the initial analysis, which looked at all the pieces of a problem such as its processes, organizations, and systems, the goal of this phase is to integrate a solution across all these areas. For example, common considerations that stretch across these three R's include the human resource requirements for the new business. This includes skills, staffing levels, and structures.

Remember, this book will guide you from the beginning of a reengineering project until wide-scale change is implemented. Not every chapter or tool described within may be relevant to your particular problem. However, you should read all of the material before beginning the project, and it also should serve as a reference during the actual implementation.

Phase **I** | *Discover*

The chapters in this phase expand on two areas relevant to the early stages of a reengineering project: personnel and scope. Before producing a detailed analysis and innovative designs, you must identify the people who will play a role in the project and define their roles. Then, you must identify those areas of a business that should be reengineered and define the specific boundaries of the area you have selected.

Let us return to the plight of Wile E. Coyote. How many times has he failed dismally in capturing the roadrunner? He fails every time, yet he continues to try. Mr. Coyote thinks about his problem and decides that he is failing because he is trying to catch the roadrunner all by himself. He heeds the maxim that "you can't do everything yourself" and decides to assemble a team of coyotes to help him achieve his goal. So, Mr. Coyote goes out and handpicks the best and brightest coyotes he can find to help him. They all assemble and begin anew to try to catch the roadrunner. Still they fail. A baffled Mr. Coyote cannot understand what has happened. After all, he got the best and brightest helpers that could be found.

The adventures of Mr. Coyote illustrate the "process" of trying to catch a roadrunner. At the beginning of each adventure, he assesses the situation. Mr. Coyote decides to use a small bit of food to lure the roadrunner to a particular spot. He figures that once he has the roadrunner stationary and contentedly eating his food, he can use one of his sophisticated traps (e.g., two-ton chunks of metal) to smash the roadrunner. Thinking that he has assessed the problem properly, Mr. Coyote goes off to implement his plan. After the trap is set up, the roadrunner runs up to the food, pecks at it, and escapes unscathed. Mr. Coyote is infuriated and perplexed over why the roadrunner's eating the food did not cause two tons of metal to come crashing down on his head. In a huff, Mr. Coyote runs out and jiggles a hidden rope connecting the food and the metal. Within seconds, the trap springs and crashes down on Mr. Coyote. Here the viewer begins to wonder about Wile E. Coyote's obsession with one particular road-runner. Time and time again the coyote plots, executes a scheme, and fails in the attempt to catch the road-runner. There are scores of roadrunners that Mr. Coyote could go after; yet he only tries for this particular one.

What do all these Wile. E. Coyote stories mean for us? Metal is falling on the coyote's head, full teams of coyotes cannot get the job done, and he has a weird obsession; but so what? These stories happen to exemplify the two areas described in the "discover" phase of this book.

Take personnel issues: leadership and team formation. Can Wile E. Coyote overcome the "leadership paradox"—that is, reengineer his coyote-catching process, given that he is the one who designed all the failed attempts? As for his team, he got the same results with them that he got working alone. What if Mr. Coyote were told that *all* coyotes are incapable of catching roadrunners? They do not have the skills for that job! Building a team is a reasonable idea, but it is *who* is selected for the team that makes the difference. Perhaps cheetahs would have been a better choice for Mr. Coyote.

Wile E. Coyote's scope selection also was faulty. Perhaps a simple trap would have done the job. Perhaps he should get out of the roadrunner business altogether. His incorrect assessment led to the application of the wrong tools to do the job. Further, investing all his resources in trying to catch a single roadrunner is an example of his having too small a scope. Expanding the scope to include more roadrunners or other types of prey probably would be more realistic.

The areas that caused Mr. Coyote severe problems may seem obvious to any manager. However, in spite of such general awareness, they often are primary reasons for failure. Therefore, before charging off to fix a business process, you must consider the plight of Mr. Coyote. Do not simply rush into a program in order to reengineer. Although this book is primarily about how to create radical change, there are four prudent actions that need to be taken up front: Chapter 2 illustrates why poor leadership is a common cause of failure in reengineering projects and gives detailed guidelines on what leaders *should* be doing. Chapter 3 outlines the different roles found in most reengineering teams as well as things you need to avoid. Chapters 4 and 5 focus on scope issues, first by selecting a candidate for reengineering and then by identifying specific boundaries of the project.

2

Reengineering Leadership

Reengineering projects will result in a less than optimal outcome, if not total failure, if there is a lack of leadership. In this chapter, we offer some insights about leadership problems on reengineering projects, and we provide specific details on what a leader should be doing. The basis for this chapter is the description of a reengineering leader taken from Hammer and Champy's *Reengineering the Corporation:*

> Most reengineering failures stem from breakdowns in leadership. Without *strong, aggressive, committed, and knowledgeable* leadership, there will be no one to persuade the barons running the functional silos within the company to subordinate the interests of their functional areas to those of the processes that cross their boundaries. No one will be able to force

changes in compensation and measurement systems, no one will be able to compel the human resources organization to redefine its job rating system. There will be no one to convince the people affected by reengineering that no alternative exists and that the results will be worth the agony of the process. (p. 107, italics added)

WHY LEADERSHIP IS SO IMPORTANT

In our opinion, there can be no stronger description than the foregoing one. The problem with a chapter on leadership, however, is that most people associate such material with lofty maxims ("fluff"). For example, we have all heard over and over again that quality improvement programs will fail unless they come from the top down. Yet leadership is important, and the poor leadership evidenced in the quality era may be followed by even worse leadership in the reengineering era. Some aspects of leadership that are germane to reengineering projects are as follows.

The Leader Does Not Need To Be the CEO

In a reengineering project, we do not expect that the leadership will come from the CEO or a senior officer. Unless the mode of the corporation is to reengineer all the key business processes or to utilize some sort of process-based structure across the entire company, the leadership must come from the executive who sits above the functional heads of the particular process scope. The leader's control must span all functions within the project scope.

Not Doing It All the Time

Reengineering leadership is temporary. Because of its disruptive nature and inherent risk, companies should *not* be reengineering all the time. The person who has to lead the reengineering project does not necessarily need to be the executive in charge of the process after reengineering is over. Although there is some overlap between what is expected of a reengineering leader and other types of leadership, we will illustrate below the leadership characteristics to get reengineering done.

All Pistons Must Work Together

Reengineering projects touch everything. Processes, systems, organizations, and compensation systems may all change simultaneously. This is a high-risk, complicated way of conducting business. What makes it really scary is that in reengineering there are an inordinate number of dependencies between project areas. In many types of projects, success can occur if the information systems people pull through, the right people are hired, or the training goes well. In contrast, reengineering is much like a car: if all the pistons are not working properly, the car will not run. Therefore, the leader must have broad knowledge in order to understand all the important facets of the reengineering project.

Managing the Chaos

Reengineering is a chaotic activity; and usually creative, talented people are more likely to work well in a chaotic environment than are other corporate citizens. Because a successful reengineering project will have a slew of these types of people, someone needs to make sense of it all, to manage it. In *The Fifth Discipline,* Senge accurately describes the peril of empowering people without clear direction and leadership. The "get it done" nature of reengineering demands empowerment. Hence, the deadly mix is there— empowered people in a highly chaotic environment. Without being a control freak, the reengineering leader must make sure that the project does not become so chaotic that its vision gets lost.

WHY SO MANY LEADERSHIP PROBLEMS?

Several factors contribute to the leadership problems that plague reengineering projects, particularly the so-called paradox of leadership. What are we asking a senior manager to do? In many cases, the answer is to somehow get creative and smart enough to fix something that management could not make work in the first place. Hammer and Champy say that reengineering is about starting over, not fixing. However, most companies would sooner fix an existing problem than take on a reengineering project to completely restructure their operations or to improve on an already world-class process.

Reengineering projects will struggle if the person who was the leader of the old broken process becomes the de facto leader of the reengineering project simply because he or she sits at the top of the hierarchy. In such cases, the leader should come from outside the process being reengineered.

Companies must be careful in starting reengineering projects. By now, practically every manager in America has heard about reengineering, and more and more companies will be engaging in reengineering, creating many new opportunities for leadership vacuums. Although leadership requires a senior-level manager, many reengineering projects begin because some smart people at the lower and middle ranks have caught on to the idea, have seen some major opportunities for its application, and have done just enough to convince the boss (and his or her bosses) that reengineering is a good thing to try. This is a recipe for disaster. Although lower-level employees should not be discouraged from promoting new ideas, it is important that their message be clearly communicated to upper management. The upper-level managers, not the people who thought it was a good idea, must take the initiative to continue the reengineering project. There is no telling exactly what would happen to renegade reengineering projects; but it is safe to say that if a high-level leader were not present, the actual results would fall far short of those described in the "sales" presentation that got the effort going.

THE CHARACTERISTICS OF A LEADER

Let us expand on the ingredients that make the ideal reengineering leader. In a nutshell, the leader must be aggressive, driven, and hard-nosed, and must occupy a high level in the organizational hierarchy. We now discuss Hammer and Champy's key characteristics one by one.

Strength

Few members of the upper management of most companies have reached their position because of their competence and hard-driving, uncompromising desire to do things right. In many cases, their climbing the corporate ladder may have resulted from consensus building, networking, and an ability to make any project seem like a success. The leader of a reengineering project must have the qualities that one would expect all senior managers to have, rather than those that they may have used to attain their positions. The best way to show where strength is needed is to look at some of the many challenges in reengineering that require it.

Fight the Good Fight

Many people will distrust the reengineering team. People at all levels of organizations involved in the process will claim that things are not really what the reengineers say they are. Even if they admit that conditions are pretty bad, they will say that the solution proposed will never work and/or is not worth the investment. Over time, if the leader does not effectively show his or her strength, some of these naysayers will ultimately sabotage the effort. The insidious little problem is that the naysayers may not just be at the "disempowered" levels of the organization; they can be found at any level. The more senior their level, the more they can be a thorn in the side of progress.

Keep Down the Numbers

On the other side of the naysayer coin is what can be termed the bandwagon clan. The bandwagon clan consists of all those managers who suddenly want a piece of the project because they feel that the input of their group is necessary to getting the job done. A colleague of ours described the phenomenon in this way: Think about a small campfire burning. The fire is nice, and a few people are sitting around it roasting some marshmallows. Suddenly the party is interrupted by several different individuals who want to join it. Having a few more marshmallow roasters is not desirable, but they could be accommodated. The only problem is that these people are not interested in cooking some marshmallows over the campfire; they have come to roast their pig. Everyone knows that one pig will extinguish a small fire used to roast marshmallows, and if there are several pigs, it will be impossible to start a new one. The reengineering leader must keep the pigs away from the campfire. It takes strength to do this, as the leader simply must keep the project small and keep un-needed parties away even though they think they should be involved.

Built-in Bullet-proof Vests
Besides disposing of naysayers and protecting the campfire, the leader must be thick-skinned. If the reengi-neering project is broad enough and is trying to break down a very traditional infrastructure, the leader must be prepared to take more than a couple of bullets. Reengi-neering leaders will be challenged at every turn, as any projects that do not preserve the status quo receive unusual scrutiny. Every other project could have failed miserably for the last ten years with no one having to account for it; but when a reengineering project is undertaken, the first inkling of trouble brings out the sharks. There will inevi-tably be problems with implementing radical change, so the leader must be prepared to deal with an unusual amount of flak.

We Will Overcome
Finally, the reengineering leader will show strength through self-confidence. The leader will be comfortable enough to listen to and take direction from members of the reengineering team who are many layers down the organizational chart. Reengineering is highly operational, and the insight of those close to the action is more likely to be on target than that of someone at some distance from it. Listening and taking direction are measures of strength, as most senior managers will have to fight themselves to let this happen.

Aggression

The difference between being aggressive and showing strength is rather like the difference between defense and offense. Reengineering leaders show their strength by defensive maneuvers—dealing with problems, holding the fort. At the same time, reengineering leaders must go on the offense. They must take proactive measures to ensure that the project will be successful.

Find the Best
The first thing on the agenda is for the leader to make sure that the best people are being used for the reengineering project. A reengineering project may grow to include as many as fifty people to implement systems and process changes. The leader must understand that it is important to seek the best people to fill the critical assignments. For example, if the process redesign calls for the adjustment of job titles for union workers, it is important for the leader to find the best person to handle that job, whether or not that person is part of the current organization.

Communicate, Communicate . . .
The reengineering leader must aggressively communicate the importance of the reengineering effort, what it is going to do, and when it will be implemented. As we will explain in Chapter 3, there is a role for a permanent public relations person in the reengineering team. The communications role of the PR person is different from that of the reengineering leader in two significant ways. First, the reengineering leader has many other responsibilities besides communicating what the project is about. Second, and more important, the reengineering leader is responsible for communications to upper management. Upper management has different concerns from those of middle management and nonmanagement employees, and the com-

munications plan that the reengineering leader uses must serve those special concerns. The most striking difference is in the power that upper management may wield. Obstacles at the lower levels usually involve larger numbers of nonpowerful people, whereas just one or two obstacles at the upper management level can cause the entire project to fail.

Steer the Ship

Pushing for results is an important role of the reengineering leader. Although the entire implementation may take several years, it is important that clear milestones be established for interim results that do not compromise the final solution. To make this happen, the reengineering leader must push the reengineering team to meet the interim milestones. Often a team spends all its energy moving toward the final solution, and it is up to the leader to make sure that the focus on early results is not lost. The same type of aggressiveness is needed to ensure that the schedule is met. In many projects, the business objectives have become dim memories by the time of implementation. The leader must constantly reinforce these interim objectives to keep the team motivated.

Commitment

Make It a Job

Hammer emphasizes that reengineering will fail if it becomes a staff function. If a few executives get together once a month to discuss how reengineering is going, the project is doomed. The leader has to be committed to the effort. We disagree with Hammer and Champy on their assertion that the reengineering leader does not have to be involved in a near full-time capacity. The more time that an executive can give a reengineering project, the quicker it will be completed and the better the results will be, especially if the leader has the characteristics discussed in this chapter.

Burning Desire

Unfortunately, having the right skills and spending a lot of time on it may not be enough. The leader must become emotionally involved in reengineering. The leader's passion should border on fanaticism. A leader who becomes closely associated with a project is unlikely to give up or become disinterested in it. In one reengineering project we studied, a leader was so dedicated that he took a crash-and-burn stance about the project, making the personal sacrifice of leaving the company after the reengineering was complete. It was not that he had no role in the company's future, but rather that he had created a very high level of discomfort within the company as the project went on. This leader was determined to make the project successful regardless of its cost to him. Although this is an extreme example, it does illustrate the type of passion that the leader must have.

In for the Long Haul

Reengineering projects do not last as long as successful marriages or even an evening law school program; but while they are going on, they seem to last forever. In *Process Innovation,* Davenport notes that many dramatic-change programs have taken up to five years to implement. We think that companies are getting much better at reengineering so that five years is too long. Nevertheless, it is still safe to say that any reengineering worth its weight will probably take at least two years to become fully operational. The leader must be willing to grind it out from the beginning to the end. A change of leadership in the middle of the project is not a good sign unless the leader was not getting the job done. Whether or not the leader is ready to stay for the long haul is indicative of how much passion is really there. Reengineering leaders cannot be as flip as those people who go crazy over the latest multilevel marketing program for a week or a month. Be wary of people who pick up all the latest management crazes and get behind them for a while.

When the Going Gets Tough ...

A reengineering leader does not give up. There will be many times when the effort may not seem worth it. Reengineering definitely has its masochistic aspects. During these rough times, the leader must stay superfocused. If not, the troops will probably give up or jump ship, and the naysayers will be like sharks around a fresh bucket of chum.

Remember Why

Ensuring that the scope is adhered to is an important function. The reengineering team members will get distracted. They may try to jump in and make interim fixes that do not move the project closer to the goal, or they may want to go well beyond the chosen scope. These distractions must be tempered by leadership. The project took a size, shape, and form for a purpose. The leader must ensure that the deck of cards is not shuffled arbitrarily by outside influences that would change the direction of the project.

Knowledge

Leaders on reengineering projects must have more than a clue about details; they must have knowledge of details. The details take several forms.

Have a Firm Grip on Reengineering

Hammer and Champy correctly point out that the leader of a reengineering project must understand what reengineering is. Because reengineering is still a relatively new field, the senior executives are highly unlikely to have any direct experience with it. We suggest that the leaders read everything available on the topic and seek formal training where some problem-solving exercises are employed.

Understanding the tools that are required to make reengineering a reality is important. However, the emphasis here for the leader is on the knowledge of what reengineering demands. The discussion in this chapter illustrates just how hard reengineering will be, and it is important that the leader know that land mines exist before they just pop up.

What Is the Blueprint?

The reengineering leader must know about the process at a more than cursory level. This will require a big jump for most executives, who usually are focused on setting a high-level direction and the financial aspects of the business. Because reengineering is a highly operational activity, the leader will have to understand what the process does, why it does it, and what its value-added outputs are.

The leader's major emphasis must be on the redesigned process. Redesigns always demand that changes in policy take place, and such policy changes may take the form of new compensation plans, new supplier management policies, and different organizational structures. We have explained the importance of the leader's breaking down barriers to make these things happen. However, it is equally necessary that the leader understand why it is important to make these changes, which are fundamental to the way the process will behave in the future. The senior management must have a firm grasp on why a new direction was set, the benefits of doing business the new way, the relationships of each critical decision with the other, and the downsides of each decision. Without such knowledge, a solution still may be implemented, but a huge gap will exist, and over time the business process will deteriorate.

The Yellow Brick Road

The leader needs to understand the mechanics of how the reengineering project is being done. This book is designed to lay out important concepts for accomplishing reengineering. Depending on the project, some or all the tools discussed here may need to be utilized. The leader must know why some things are being used and others are not. More important, however, the leader needs to know why these tools are vital to the effort, including what they are and what they are designed to do. For example, the leader should have enough knowledge about process simulation to know not only that it is being done but that it is important for validating the redesign, staffing, and so on. There are two reasons why such in-depth knowledge is necessary. First, the leader will be forced to make many presentations about the project, and it is always extremely impressive to employees when the person running the show seems to know what is going on and especially why it is going on. Second, reengineering is temporary, but the tools discussed in this book that are used to accomplish reengineering are not temporary; and the senior management must know that the tools are reusable corporate assets that can be applied to many projects (e.g., code generators) or ingrained in the way that business is done (e.g., training paradigms).

To summarize, reengineering leadership may be best explained by an old adage from the legal field: substance always outweighs form. In law, whenever a lower court's decision seems to turn on a technicality, inevitably the upper court finds a way to ensure that the substantive issues control the outcome. In reengineering projects, executives who are flashy or politically astute are just form; only leaders who have a firm grasp of the spirit of the reengineering project and a deep understanding of what the project is designed to accomplish will take reengineering from a corporate agenda item to corporate reality.

3

Selecting the Reengineering Team

The presence of a strong executive leader who is committed to the project does not diminish the need for a well-balanced team responsible for implementing it. Executive leadership is necessary but not sufficient for a reengineering project to succeed. New York Yankees Manager Casey Stengel recognized this when he said, "I just know I'm a better manager when I have Joe DiMaggio in center field." Dealing with a myriad of technological and human resource issues will require the work of many people at different levels in the company. Thus, once there is sponsorship for the project, the most important question is "Whom do you select to be on the reengineering team?" The particular personnel who are chosen and the amount of authority they are given are critical factors in the success of the reengineering project (and, indeed, for most task-focused teams).

The successful applications of concurrent engineering illustrate the significant gains that can be realized by forming cross-functional teams.[1] Because the very nature of reengineering projects is cross-functional, this paradigm bodes well for reengineering teams. However, the reengineering "core" team must start small and remain small. Therefore, the prudent approach is to find as few people with as many of the requisite cross-functional skills as you can. Five to seven individuals should comprise the core team. There may be cases where fewer than five people can be used, but always avoid teams of ten or more. Ten people rarely agree on anything, and building the required synergy with ten people is too difficult in a short period of time.

Because the reengineering team must be cross-functional, a balance of outsiders and insiders must be selected as full-time, dedicated team members. Outsiders are people who have not been involved with the affected organizations, and insiders are people who have had recent experience as a member of one or more of those organizations. The core team members must be complemented by two other individuals: a public relations person and a champion.

We must note that, in addition to their skills, the way that the team members are perceived is important. How the affected organizations will view each member of the team is a factor in team selection. The goal is to select team members who have the requisite skills and also will be able to gain some acceptance from others in the current environment.

In discussing perceptions and prejudices within a system of jurisprudence, the notable Supreme Court Justice Cardozo wrote, "The forces of which Judges . . . avail to shape the form and the content of their judgments" include the "likes and dislikes, the predilections and the prejudices, the complex of instincts and emotions and habits and

---

[1] In a survey of concurrent engineering projects coauthored by Dr. Jim Pennell (see the Suggested Reading), reported benefits of concurrent engineering included reduction of time to develop a product by as much as 40 to 60%, the reduction of manufacturing costs by as much as 30 to 40%, and the reduction of rework by as much as 75%.

convictions, which make the man, whether he be litigant or judge."[2] Because reengineering projects are usually high-profile endeavors, the team members will be judged constantly by the people within the organizations. Therefore, the selection of team members must be done so as to avoid negative personal perceptions. For example, negative references to "MBA know-nothings" must be avoided in the selection of outsiders. The insiders must be individuals respected by most people at all levels of the organization from which they came. If Bill is viewed as a management toady or Sue is generally seen as incompetent, then the team will be handicapped at the outset. If they truly are the best people for the job, then you must determine why they are perceived negatively, and you must take steps to overcome those perceptions.

THE REENGINEERS

REENGINEER WANTED

We at XYCO Technologies are seeking a candidate to assist in reengineering our accounts payable processes. Applicants for this position should have strong analytical *and* creative capabilities. The successful candidate will have a firm grasp of information technology and quality improvement methods. The project demands a thorough knowledge of what makes a business more productive and a willingness to work at all levels of the company to achieve the project objectives. Moreover, he/she should be confident and self-motivated, and have strong interpersonal skills and a passion for doing things right.

---

[2]Supreme Court Justice Benjamin Cardozo in *The Nature of the Judicial Process,* Yale University, 1921, p. 167.

This may sound like a pretty tall order, but all these attributes will be required during the course of the reengineering project.

Although the prototypical reengineers will be "management"-level persons, they must not be responsible for managing people during the reengineering project. They need the time and the freedom to analyze and diagnose problems. They also must be able to digest minute details because the people working in and managing the current environment generally speak about details. When reengineers "know" the details of the current environment, the employees working in the process are more confident in them, and that attitude inspires employee confidence in the overall effort. At the same time, however, the reengineers need to be able to see the "big picture." Peter Senge calls the talent of understanding the interrelationships of fragmented information sources "systems thinking." This talent is critical for the reengineer so that the fragmented pieces of the process can be reassembled to create a complete picture.

An understanding of technology understates the level of expertise required in reengineering. Here the people must be technically *complete*. The more experience they have in information system integration, the better. Prerequisites include a thorough understanding of information/ data analysis, database technology, and means of exchanging data. Finally, the more the reengineer knows and understands Total Quality Management methods, the better. As described in the "hunt and gather" phase, these methods will be useful for identifying the root causes of problems and establishing continuous improvement programs after the project is complete.

The "hard" skills mentioned above must be complemented by a wide array of "soft" skills. The reengineer must be able to go to the factory floor and relate to the workers and then must be able to go to the boardroom and present the team's findings in a way that the senior managers can understand. Similarly, reengineers must be able to infiltrate situations. This may sound like a negative term, but getting behind the scenes is a skill that helps one to find out what is really happening.

In addition, reengineers tend to be nonconformists. However, ideas that challenge the status quo usually come under scrutiny and attack; so these nonconformists must be thick-skinned. The reengineers must be able to defend their ideas and conclusions.

Self-motivation or "drive" is a characteristic of the people that you want involved in your reengineering project. At the beginning of such projects, there is typically a lack of specific direction because the process of "looking to create radical change" is a highly creative one. Given this milieu, the members of the team are expected to and must create opportunities.

Finally, because reengineering is not a paper exercise, the reengineers must gain the insight necessary to make good decisions on a process redesign idea and/or a new information technology solution by spending a great deal of time getting their hands dirty. Ways to understand the current environment include interviewing, observation, and literally working the process oneself.

Clearly, the profile of a reengineer is a little more complicated than that of someone with good technical and interpersonal skills. Where do we find such people? The first place to look is in an R&D organization or an "R&D-like" organization. Even though R&D types are not the prototypical management consultants who bring in the latest management techniques, we have noticed that a large number of systems professionals and engineers are more interested than ever in how technologies impact their business. Also, keep in mind that such management gurus as Deming, Juran, and more recently Hammer and Senge certainly would be characterized as R&D types. A second-

ary source would be people who left the R&D environment expecting to become the new wave of senior managers.
Many of these people are frustrated with the traditional
side of their business and welcome the opportunity to
pursue new challenges. These people also may be a better
match than the traditional R&D type, for not only do they
have the needed technical background but they also have
recent business experience. In both cases, however, the
reengineers have a technology background. We strongly
believe it is easier to teach process analysis skills to engineers and computer scientists than it is to teach technology skills to the typical manager. Finally, some of the
reengineers can be external consultants. Because the reengineering discipline is relatively new, augmenting the
team with consultants will help to speed things along and
provide needed experience. However, the role of the consultants must be clearly identified. Consultants should be
used to help teach the reengineering team members techniques, provide objectivity, and steer the team around
roadblocks. The consultants should not be solely responsible for all aspects of the project, nor should they merely
provide a report to management.

INSIDERS

At least 30 percent of the reengineering team should be
comprised of employees who have recently worked in the
current environment. In dealing with the potential perception problems discussed earlier, there may be no quicker
way to gain credibility than to have highly respected
members of the current organization involved in the effort.
We have seen teams consisting solely of outsiders and
have watched them fail.

When the insiders are selected, be cognizant of the
three levels of commitment that Peter Senge discusses in

The Fifth Discipline. In a discussion of developing shared visions, Senge points out that people may have several reactions to change programs such as reengineering. The first of these is enrollment, which is symptomatic of their being "on board" or "part of the team." Second is commitment, which usually is expressed in "wanting" change without having the do-or-die attitude needed to make it happen. Third, there is complacency; those people who are complying are doing so to go along for the ride or perhaps in order not to rock the boat. Finally, there are those people who will not comply or may just be plainly apathetic. An understanding of these attitudes ensures that the right people from the current business environment are selected for the reengineering team. At the same time, the insiders who are chosen should be able to influence their former peers and associates by carefully evaluating where they fall on this scale. Common sense tells us that the chances of successfully reengineering a business are improved if more people are enthusiastic about the opportunities created by the reengineering program.

THE PUBLIC RELATIONS PERSON

To complement the core team members, it is necessary to have a PR person on the reengineering team. This person should be dedicated to spreading a positive message about reengineering and the benefits of the program in particular. This is more than a cheerleading exercise; it is a concise, credible dialogue with employees, discussing the findings of the reengineering team and their plan for the future. The PR person must be out in front of the team, talking about the reengineering project and listening to employees' concerns. This person can help smooth over the inevitable perception problems about what the effort really is trying to accomplish. In simple terms, the PR person is trying to create positive associations with the project in people's minds. The targeted audience needs to understand what the project's advantages are and how it will either benefit them in the future or, at a minimum, reduce some of their current frustrations. The PR person must construct the communications program carefully. Because the ultimate impacts of the reengineering program on issues such as staff reduction may not be known early in the project, the initial communications effort must focus on building a rapport with the staff in the existing environment. The PR person must be able to touch on the issues that are of concern to the employees. (These "hot spots" are generally found in employee attitude surveys, as discussed in Chapter 4.) An explanation of how the reengineering effort will address it is required for each of those concerns.

The PR person's communications plan must focus on what the positive outcomes of reengineering will be; that person must break down the emotional barriers to reengineering. No matter how good the redesign ideas are, or the level of commitment by senior management, there will be major problems during the implementation phase if the employees are not at all receptive to the reengineering team and its work. A large number of disgruntled employees can create a level of cynicism that may cause long-term harm.

THE CHAMPION

In Chapter 2, we described the importance of senior-level leadership in reengineering efforts. Within the context of leadership, a "champion" may be the most important cog in the reengineering team. In some cases, the reengineering leader and the champion may be the same person; or, in larger efforts, the champion may report to the leader. This person will have the same traits as the prototypical reengineer that we described earlier: drive, motivation, tremendous interpersonal skills, and so on. The champion is responsible for leading the reengineering team, fighting political battles, removing obstacles, and gathering support from all levels of management.

Often the reengineering team members assume that one of them can fill the role of the champion. After all, they are the people who are most familiar with the problem and have the skill and the background for dealing with it. However, the selection or the emergence of a champion must come from a higher organizational level than that of the people on the team. It is not that one of the team members is incapable of assuming this role; but unfortunately one's level in the organizational hierarchy is still paramount in many companies, and so the leader or champion must come from a high-enough level to get the attention and the respect of the employees.

Hopefully, the champion will be chosen by the reengineering leader or will emerge on the scene by him- or herself. If you are the leader of a reengineering effort, note that a recruitment process may be required for this position, and then the person must be "sold" to the rest of the reengineering team. On the other hand, if the project is beginning in a more bottom–up fashion, a nomination process may be required. In the latter case, the reengineering team must sell senior management on the team's need for this person. Settling for someone merely for organizational convenience is not appropriate and can be disastrous.

If the reengineering team is involved in the selection or the nomination process, it must exercise some caution. Consider this scenario: The process problems have been identified, and the team members have begun to look toward the horizon. They come up with some ideas on what and how to reengineer. The team members usually will begin to share their ideas within the circle of their immediate acquaintances, which may include their division manager or their immediate supervisor. The trap is that the team may falsely believe that one of these "initial" contacts should be the champion. This is a big mistake. Instead, spend a month or so spreading the word in a larger circle, explaining the process problems and some of the fundamental reengineering concepts. Then, take a fresh look at this audience—their reactions, perceptions, excitement, and potential motivations. Only then is it time to select or recommend a champion.

ORGANIZATIONAL REPORTING

As soon as the initial team is chosen, you should carefully consider the team's organizational reporting structure. Try to avoid a matrix management reporting structure. Remember that the focus was to form a cross-functional team, not a cross-organizational team. The drawback of having the team members report to different functional managers or all report to one functional manager (other than some senior reengineering leader) is largely one of perception. If people continue to report directly to their original manager, and in addition report to another functional manager, the perception is that the organization offering the team members' participation is only giving lip service to the effort. If the members of the team all report to one functional manager, the team will be perceived as a "brain-child" of one organization. Once this happens, resistance definitely will form, not only in the operational environment that is being reengineered but also within the other functions whose resources were tapped to support the effort.

Ideally, the team should report directly to the executive sponsor, as this would show that the sponsor was very serious. Although the champion will be the operational leader of the team, the team needs direct access to the reengineering leader. We recommend that a separate organizational structure be formed, but this organization should be considered temporary. The separate group must not be considered a permanent quality department or process management function. It is important for the reengineering group to avoid all the baggage that goes along with traditional organizations and to remain a special assignments group. Figures 3-1 and 3-2 illustrate team members' reporting structures before and after they were formed into a reengineering team.

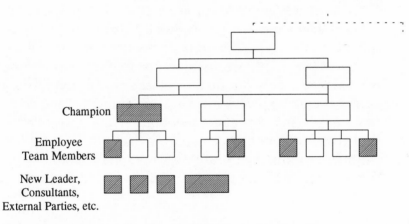

Figure 3-1. Organizational structure prior to forming the reengineering team.

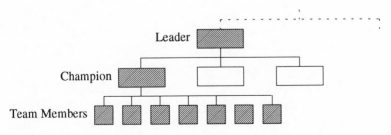

Figure 3-2. One possible team reporting structure.

THINGS TO WATCH OUT FOR

The reengineering team must always be a cross-functional team. Cross-functional must be distinguished from cross-organizational. At no time should the team be built with the premise of "getting people together from each of the organizations" to work on a problem. This paradigm will not yield breakthrough solutions. Instead, it encourages the type of thinking that already exists: Paul is representing systems, Mary is representing billing, and so forth, and each individual focuses through the eyes of his or her current organization.

Avoid compromises in selecting team members. We have seen reengineering projects where an ad hoc selection of team members permitted a "bad apple" to be placed on the team. The drain on the team was immense.

Avoid keeping the team intact just for the sake of doing so. No matter how careful the personnel selection process is, there is some probability that one or more of the original team members will not work out. In this case, bite the bullet and remove that person from the team. Failure to do so may result in significant degradation of productivity, which can hardly be afforded in time-critical projects.

4

Business
Assessment

The purpose of this chapter is to provide
a means of identifying problem areas in your business
and using those areas to set the scope of the reengineering
project. In our experience, seven categories are useful in
assessing the business. For each category, you should
compare the description in this chapter (both the quantitative and the qualitative aspects) with the corresponding
area in your own business. Taken together, the results of
all seven comparisons will give you a gestalt sense of your
business and how it compares to businesses in need of
reengineering. Further, this material will help you focus
on setting the appropriate scope in the next chapter.

Although you will be gathering some data about your
processes in this chapter, the intent is not to compile all
available process metrics at this point. Much of what we
have seen in the way of measurement parallels the findings of Deming. That is, management most often measures

the wrong things and uses the data gathered incorrectly. At the beginning of a reengineering project, you should avoid detailed quantitative measurements. For example, if you are thinking about improving a customer complaint process, then you do not need to consult the last week's financial data or find out the number of complaints last month in order to carry out the initial business assessment. Instead, we recommend a subjective focus on the seven process categories. At this stage of the project, you are really looking for a subjective or "gut" view of the problems with the business. Carrying out the business assessment and setting the scope should take no more than a month. Later, in the "hunt and gather" phase, a more in-depth analysis period will be necessary.

PROCESS MEASUREMENT CATEGORIES

The process categories, listed below, are interrelated. They are described in detail in the following sections and then are compiled into a table near the end of the chapter. The beauty of using the assessment table is that it provides a quick snapshot of how the business is performing. The categories are:

- Personnel cost trends
- Computer system cost trends
- Throughput
- Hand-offs
- Customer satisfaction
- Employee satisfaction
- Management attitude

Personnel and Computer System Cost Trends

Two areas that can quickly be tracked back to annual budgets are cost trends in human operating costs (head count) and cost trends in computer system support. If the business process is supporting a new initiative (e.g., a new service offering) and substantial ramp-up was required, the ramp-up years should not be considered here. Once the ramp-up period has been eliminated, several questions must be asked. Has the number of people required to perform the process been growing, stayed the same, or shrunk over the last three to five years? Ask the same question about hardware and software costs. The purpose of asking these questions is mostly to get a handle on spending trends. Many times a 5 to 10 percent annual increase goes unrecognized because when it occurred, it was dealt with as an aberration. This is especially true when new leadership has taken over somewhere in the middle of the measurement period. A 5 percent annual increase over five years in either of these areas points to a serious problem. This is true even when there is more business and revenues.

Let us take a common-sense view of the situation. Assume that your business process is the insurance claims process. If there were ten insurance claims adjusters in 1989 and there was a mere 5 percent annual increase in staff over five years, there would be thirteen people running the process in 1994. That does not seem so bad. However, what if you started with 100 people? After five years at the same rate you would have close to 130 people, and thirty people can cost more than three million dollars annually. Management may notice this situation when profits are down, but this problem is most insidious when the business seems to be doing well financially. There are usually one or more root causes of this problem:

1. There is empire building going on. As a staff's size increases, the "boss" of that staff has more power. The reaction is that John wants to add three people so I certainly need three more.

2. If there is more work to do (i.e., business is getting better), management may throw people at the problem. This approach is usually meant to be temporary, but temporary staff tends to become permanent.

3. Quality problems and rework may result in backlogs, and instead of trying to improve the process with the current staff, management may attempt a quick fix by adding staff.

Organizational growth is cancerous if it cannot be supported *in the long term*. Thus, a trend of increasing personnel costs may indicate the need to reengineer the business.

The same is true for trends in computer system costs. Many business processes are directly supported by information systems. This is especially true in the more traditional operations processes such as billing resolutions, order handling, and so on. The idea of assessing computer system support costs parallels the assessment of human resource costs in several ways. First, computer systems are supported by the people (typically in the management information systems [MIS] organization) who are responsible for software development as well as the production support personnel. Production support consists of the staff designated to fix information system environment problems such as network congestion, application errors, and database problems; they are responsible for the day-to-day maintenance of the information systems. The required MIS staff equate to human resource costs. When looking at the process costs and trends, be sure to include the information system head count. A second computer system cost area is the physical hardware and software costs trends, which include the hardware and software upgrades required for systems to stay up to date or to increase capacity. If these annual costs have been increasing steadily, there usually is a problem. Some cases, such as mere memory upgrades on a particular processor, usually are not problems because the costs of increasing memory usually are smaller than the human resource costs required to achieve the same gain in performance. We are concerned here with unanticipated hardware upgrades that constitute more than 5 percent of the total yearly budget. These costs occur when there are system performance problems due to business increases that cripple the information system, or when the hardware platform is so poorly designed that it is totally dependent on a multitude of unrelated software products that require simultaneous upgrades to keep the environment working. These symptoms are further explored when we discuss how to uncover the root causes of the information architecture problems in Chapter 8.

Throughput and Hand-offs

Webster's simply defines throughput as "output" or "production." Why would one want to examine the output of a process? Typically, processes are measured in terms of cycle time (how long it takes to get from point *A* to point *Z*) or defect rates (how many widgets are broken). These are important measurements for statistical process control. However, in a high-level evaluation, they may be expensive to procure as well as misleading—they are really "in-process" measurements. Used in combination with operational cost trends, the throughput can give a quick high-level understanding of how a process really is performing. You are trying to find out if your process is uniformly creating the same amount of product regardless of conditions, which include input rates and operational cost trends such as systems and staff costs. There are several scenarios to watch out for. For example, the output of the process is highly sensitive to the input rate. If you are receiving ten claims a day and you are outputting ten, then the process is holding a steady state. If the process is receiving ten claims a day and can only output two, then common sense tells you that if this keeps up, the process will become saturated with claims. This small example of a general principle should give you an idea of what we are talking about.

Instead of looking at short-term measurements, you need a long-term view of the process. Like the trends in operating costs, throughput should be evaluated over several years. If the process is claims resolution at an insurance company, how many claims were received last year? The year before? How many were completed last year and the year before? Now, what were the operating cost trends doing at the same time? If the output rate is not keeping up with the input rate and the expenses are growing, the situation is becoming serious. Again, this analysis is not meant to be a thorough statistical evaluation. If the process truly needs a serious overhaul, some of the numbers discussed so far will jump off of the page at you.

Our definition of a hand-off is the complete relinquishing of an unfinished product by one worker (or group of workers) to another worker (or group). Frequently the remnants of functional processes invented in the 1950s, hand-offs can be the most insidious problems in a process. Work was done then by functional organizations (e.g., engineering, manufacturing), and within the organization there usually were further breakdowns of work by specific work titles. It is safe to say that if there are too many hand-offs in a process, the process is in serious trouble. Hand-offs automatically mean that time, accountability, and empowerment are being lost. The key question is, how many hand-offs are too many? Unfortunately, there are no magic answers, but there are some rules of thumb. However, any hand-off at all should be considered suspect. Several strategies for eliminating hand-offs, such as team approaches and case workers, are discussed in subsequent chapters.

Customer and Employee Satisfaction

These two indicators usually go hand-in-hand. If customers are not happy, employees generally are not happy (either because the customers are breathing down their necks or because the employees are personally dissatisfied with the company's inability to serve the customers). Similarly, if the employees are unhappy because of poor wages, overbearing management, or lack of direction, those frustrations often are evident in their interactions with customers. Much has been written on the topic of corporate cultures and the problems and the benefits that can be lost or gained depending on the "health" of the organization. Several examples included in *The Deming Management Method* by Mary Walton and *Principle-Centered Leadership* by Stephen Covey describe the relationship between employee satisfaction created through participation and the general financial health of the organization. The best way to find out if customer and employee dissatisfaction may require an overhaul of the business is to look at these areas over the last few years.

One nicety that emerged from the quality era is that companies now recognize that these areas are important to their success. Therefore, most organizations within companies have adequate ways to determine how happy or unhappy the employees and customers are. There probably are historical data for the last several years that measure these areas. Consult these data, and make a determination about your own situation. It is unlikely that other indicators will point to disaster and this one will be rosy; so, if that is the case, you may want to review your assessment of other areas. Have your scores in this area been consistently low? If so, it is fairly safe to say that there is serious trouble within the process. This is especially true if the organization has been involved in quality initiatives over the last few years. A good quality improvement program with committed management will yield positive results in this area unless the process needs an overhaul. If your review shows that things seem to be getting better, you may want to wait another year to see if that trend continues. If you find that the situation is getting worse, then you are in a gray area where a combination of other measurements is important.

A word of caution. Do not automatically dismiss the notion of totally revamping the process under the theory that downsizing and churn will only make matters worse here. First, there are ways to minimize this risk, as discussed in Chapter 16. Second, this may be your last chance to make the changes needed to keep the company or the organization in business. Do not miss this opportunity by being overly cautious and not wanting to rock an already unsteady boat. If the whole organization caves in, delaying a radical improvement initiative will be more detrimental to the mental health of the people in the organization than would a proactive approach for strategic downsizing.

Management Attitude

Books written on the topic of quality improvement stress the importance of top-level management support for quality programs. Time and time again they say that if the top-level management does not really buy in and participate, the long-term cultural changes required for these programs to work will not occur. As noted in Chapter 2, many of the same things have been written about reengineering.

There are critical differences in the management attitude required to successfully implement a reengineering project versus a quality improvement program. Quality improvement is a "nice" thing. It is primarily a nonthreatening discipline. The quality programs stress employee empowerment, open-door policies, and lifetime employment. Learning to trust is at the foundation. This makes the transition between the strict, hierarchical, control environment and the new environment brought about by quality largely psychological. These programs usually do not involve any specialized skills. Employees are given greater freedom. Managers are taught how to build teams and to be facilitators. Although there is often a push for additional training that will help employees function better in the core competencies of the business, the implementation of a quality plan does not require this knowledge. It is in a sense a management technique. This model contrasts sharply with the basic paradigms of reengineering, as reengineering demands a complete reevaluation of everything. Not one person involved in a reengineering project and not one person potentially impacted by it will ever feel comfortable with what is happening. For as happy and spiritual as quality improvement can be, reengineering can be unhappy and threatening. As far as management's attitude is concerned, the important question to ask is whether the powers that be really want significant changes to take place. The management attitude across the business must be one almost of desperation.

Table 4-1 summarizes the seven categories and, for each

Table 4-1. Business assessment table

Categories	Business is in a steady state	Business is in need of reengineering
Personnel cost trends	Over the last several years, costs have been pretty static. There is neither a dramatic increase in overall head count in the area nor a reduction.	Over the last several years, there has been a steady or significant increase in the amount of people required to run the operation.
Computer system cost trends	Over the last several years, the head count for computer support has remained static when a maintenance mode reduction was expected.	Over the last several years, the costs in this area have increased. Hardware costs and third-party software costs on the rise.
Throughput	The process is not highly sensitive to surges. Over time the products output are steadily keeping up with the inputs.	The process(es) are very sensitive to change in the input rate. Under normal loads things are OK, but as loads increase less product comes out.
Hand-offs	There are hand-offs in the process. However, the people doing a particular unit of work are performing value-added functions requisite to their skill level.	The product created must be touched by many people in many departments. Look out for human I/O devices.

Continued

Table 4-1. *Continued*

Categories	Business is in a steady state	Business is in need of reengineering
Customer satisfaction	The scores are not the best, but the customers have not left, nor do the scores seem to be getting any worse.	The trend is that customers are very unsatisfied. Be most aware of when the perception is incompetence.
Employee satisfaction	The employees are not totally unhappy. There seems to be more room for empowerment and self-direction.	The morale is very low. Nothing that you try helps, and it all is perceived as a fad. There is poor customer interaction.
Management attitude	Management is looking for some change. Education on quality improvement has taken place. Management would consider reengineering a threat to them.	Management is ready for big change. They are fed up. This is easy to detect with new leadership or management, early in their careers.

category, provides a contrast between a company in a steady state and one in need of an overhaul.

It is not always easy to tell where in the spectrum a particular problem lies—whether it is a disaster, or it can be fixed up with a little TLC. This chapter has explored some of the areas that are important to consider before applying a fix. The most important point is to avoid overanalyzing. You could be wasting precious time looking for an exact answer, or the numbers simply could be leading you astray. In the next chapter, you will be using assessment material to help you set an appropriate scope for the project.

5

Setting the Project Scope

By their very nature, reengineering projects require an extended scope. One of the fundamental purposes of reengineering is that (unlike most process improvement initiatives) it calls for the assessment, analysis, and redesign of business activities that *span multiple functional organizations.* This concept of an extended scope must be counterbalanced with something we call "doability." Having an extended, or grandiose, scope will be for naught if it is impossible to implement in a reasonable amount of time or for a reasonable amount of investment.

Many times the ideas needed to redesign business processes are discovered in solutions that will take a company from the red to the black. A clash occurs when such a transformation is expected overnight and the redesign team estimates that it will take several years. In *Process Innovation,* Davenport's research on over 100 companies involved in reengineering showed that not one had undergone a substantial redesign in less than two years. The much publicized Ford example (where Ford reduced its accounts payable staff by 75%) took five years. The bottom line is that reengineering projects are not quick fixes. The broader their scope, the longer they probably will take. This chapter provides you with a step-by-step approach for defining the scope of your project.

Reality may be the most formidable opponent to establishing an appropriate scope. Many organizations are in desperate need of reengineering, but because they are in such dire straits, they need results quickly. The adage "I know it takes time but we don't have any" is very applicable. A reengineering project should not begin with the idea that it is going to take *years* to see any meaningful results. Instead, the realities of the business situation should be taken into consideration in almost every activity. If a project is scoped properly, it is much more likely that gains can be made in the short term, that the golden eggs will not be missed, and that the new business design idea will be achievable.

A STRATEGY TO DEFINE THE APPROPRIATE PROJECT SCOPE

At the outset of this book, we discussed the not-so-hypothetical plight of Connie Roberts. To refresh your memory, Connie headed an organization that was experiencing significant problems in the categories discussed in the business assessment section (Chapter 4)—increasing operations costs, poor quality grades from customers, morale problems, and so on. Connie's business needed to be reengineered. Before we jump into a discussion of how to select the scope, we should ponder the realities of Connie's situation—and those facing many organizations that are candidates for reengineering.

Connie Roberts's problem has been chosen as the area to be reengineered here. A business assessment has proved that this is the type of problem conducive to reengineering, and the small team is ready to nail down the scope. There are six steps to follow in setting it, as discussed in the following sections. These steps are facilitated by utilizing the measurement categories in Chapter 4.

Step 1: Define the Process Boundaries

We need an output or a product for this exercise; so without being too scientific, we will choose one! For this example, we will use books, a popular item. In Connie's example, a preliminary attempt to define process boundaries may stay in the bounds of her present organization (although it is highly unlikely that overhauling Connie's organization alone will bring lasting results). We decide that the beginning of the process is the moment that a customer's order shows up in Connie's organization. The end is when the customer's order has been given a firm date to ship. Figure 5-1 depicts a high-level view of this process with inputs and outputs.

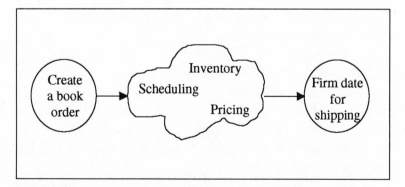

Figure 5-1. The first and last tasks in the book-ordering process.

Such identification of the beginning and the end of a process is a well-known quality technique. The quality improvement strategists go astray when they assume that the process identified makes any sense at all, or more frequently the process itself is really a subprocess. The difference here is one of degree. For example, the process that Connie is concerned with is what we just mentioned—customer order to shipping. Many quality improvement efforts will choose the process as that of getting the shipping request to the shipping department, which may seem cross-functional because the improvement involves at least two departments—shipping and order entry. However, its scope may be too small because the real problem in the larger process is that there is no way to know up front if the book is even in stock. Hence, you may do a great job of getting shipping tickets to the shipping department, but there is no inventory.

Step 2: Use the Assessment Categories

In Chapter 4, we provided a general process assessment table to help gauge the severity of the business problem that you are studying. Using the results of that initial assessment, you can take a preliminary look at the scope. For the scope selected, examine the process in terms of the following four categories.

1. *Look at the operational cost trends.* Get a ballpark dollar figure for how much more the organization is costing now than it did before for the process endpoints selected. It is also not unreasonable to use the current costs. The current cost is important for the chosen scope because once the potential scope is enlarged or shrunk, the dollar figures will be easier to match.

2. *Look at the trends for computer support for this operation.*

This is very similar to the previous category. Be sure that you are looking only at expenditures that support the part of the process you selected in Step 1. This may be difficult because many computer systems are used by multiple departments, and the costs for support and maintenance are allocated to the various user groups.

3. *Look at hand-offs.* Where does the order go? How many people touch it? Where is it being sent from one department to another via a pony express method? Is it sent to an external resource at all, such as some supplier? A very telling question is, how many different times does one person touch the same order? Are there hand-offs to rework or quality inspection groups? You get the idea—a number is needed here. Chances are that if the number is large by assessment standards, then it will not change dramatically if the scope is only slightly changed.

4. *Look at throughput.* There are at least two ways that throughput is hurting the process. In the first place, there are a lot of things being started, but very few are getting done. (This symptom is found most often in a service development process where new products are being created. There may be ten new product ideas at the beginning of the year, but only one has been completed at the end of the year. This is a throughput problem, but not the type found in traditional processes such as customer order fulfillment.) The problem that is seen over and over again is the snake swallowing the pig. The process is the snake, and the pig is a collection of orders stalled at some point in the process. The reason that this becomes a throughput problem is that there is nothing stopping the organization from taking more orders, but there are few getting out to the end of the process. (Once again, if you are considering too small a scope, there may not appear to be a throughput problem, and the presence of pigs may not be evident.)

Step 3: Analyze Other Factors Important to Scope Decisions

The business assessment certainly gives the problem a flavor. However, what those indicators tell the most is whether or not the problem domain chosen (i.e., the project scope) is large enough. Our experience is that you will have little problem finding a large scope where all these indicators seem to suggest a need for reengineering. In choosing a "doable" scope, there are difficulty factors that will determine how well you score. The project may or may not be successful, based on the choice of scope. As in diving, you must select a difficult problem in order to achieve the hoped-for results. However, choosing too difficult a problem will only lead to disaster. There are several approaches here.

1. *Analyze just pure people numbers.* As Jim Morrison says, "We've got the guns, while they got the numbers." If they have the numbers, having the top guns may not be enough. This is certainly a paradoxical situation. Having more people in an organization (so that the human resource cost trends are through the roof) indicates the need to reengineer, particularly when there are huge numbers of people. These vast numbers may seem to present easy targets because there are always gross amounts of waste in such processes. The problem is that the more people there are, the longer it will take to complete the reengineering effort. The number of people can negatively affect the effort in three ways. First, the process must be analyzed in some detail. This entails finding out what all these people are doing. Second, some people are simply obstacles, and the more people there are, the greater the chances that some of them will be obstacles to the change effort. Third, many reengineering projects result in force reductions. There are right ways to make these force reductions, which take careful planning as discussed in

Chapter 16. Obviously, it takes quite a while to manage force reductions in a practical way without destroying morale. This difficulty is only exacerbated when there are very large numbers to contend with.

2. *Analyze all the computer systems supporting the process.* Most processes are not very well supported by information technology. Moreover, a reengineering solution may require the introduction of new information technology. The worst-case scenario for a reengineering project is for all the departments or subgroups to be using their own systems to accomplish their piece of the action. This problem is similar to the problems encountered in dealing with large numbers of people. Analysis time must be spent on each system. What does it do? How does it support the process? Also, having a large number of computer systems adds complications and delays during the transition phase. Figuring out what to do with the systems or even simply retiring them is not a zero effort/zero time proposition.

3. *Discover whether there are friends or foes inside.* Often, the most appealing reengineering opportunity can be likened to the Halloween apple with a razor blade inside. From the outside, it looks good, but the first bite is full of surprises. Take Connie Roberts as an example. She may know that she needs to make radical changes and do everything in her power to get the project off the ground. However, once the team starts poking around her organization (the apple), she becomes the razor blade. This example is a subtle one that is difficult to grasp until the project is under way. Most of the time, however, the razor blades are sticking far enough out of the apple that you can see them before you bite it. The insiders identified in Chapter 3 should help in the razor blade identification. Sometimes, the razor blades can be extricated, either because there are few of them or because the management leadership is from a level high enough to take them out of the picture.

4. *Learn what the culture is like.* Cultural problems in

organizations are among those problems that quality improvement efforts aim to eliminate. The concept of empowerment is a fundamental element of these programs. The theory is that happy, healthy workers make better products and satisfy customers better. The culture needs to be evaluated slightly differently for reengineering projects. Like people, culture can be a friend or a foe. Several scenarios are found:

• *Poor culture/poor change climate:* The scenario to shy away from the most is one in which the culture of the organization is poor and also is resistant to change. If there is low morale, that is not necessarily an impediment to radical change. (In fact, some organizations have low morale, but they also seem to thrive in it.) A high turnover rate is usually a sign of "organizational problems," but the poor culture/poor change environment presents just the opposite situation. We will call it the "no turnover" organization. Organizations become stagnant when they are not injected with new blood. The lack of new people in an organization combines with other "problem factors" such as high operational costs to form an extremely intransigent situation. If the scope that you have chosen includes many organizations in this condition, you may want to reconsider the scope. Of course, the right type of executive sponsorship can break down many of the barriers; but time is always a key concern, and time spent fighting with an organization of unhappy and unfriendly people is not time well spent.

• *Poor culture/good change climate:* Often organizations with very low morale are reaching the pain threshold where they are open to change. Most often the people who are willing to change the most are the least likely candidates, as they are usually nonmanagement employees in the organization. These employees are at the front lines dealing with the poor processes that management has implemented, and they remember times when things were better. Surprisingly, the potential elimination of menial jobs is not half as scary to those doing them as is the elimination of an arcane process to the management that developed and implemented it. Finding an organization that will do anything to avoid further pain is certainly a good sign for a reengineering effort.

• *Good culture/good change climate:* There are times when you may want to reengineer to stay in front of the competition or to diversify into another business. An organization that has been successful in the past and understands the need to stay in front because of the rewards that brings is receptive to new ideas. Frequently, because of their wide scope, reengineering projects will span multiple groups within a large organization. Within these groups or departments, there may be some pockets of happy workers and good work environments. In making a decision about the scope, it may be prudent for you to include some of these groups. They tend to help with the effort because they are positive about their jobs. In addition, if they are excluded from a major effort, that may act as a demotivating force.

5. *Consider the existing experience level.* The need for experience was somewhat covered in the team selection process (Chapter 3). The actual reengineering team must be experienced in order to accomplish the difficult job ahead. How about the experience level in the current environment? Within the work groups, a total lack of experience can be an extreme impediment to progress. At the same time, as we mentioned above, complete stagnation is not very good either. The best scenario is for there to be some "experts" within the organization who really know what it takes to complete their task. This is extremely important in complicated processes that require special technical skills; it is very important that some people

know their jobs really well. A good process design must be founded on a strong understanding of the existing environment. A lack of understanding creates the possibility of re-creating some of the fundamental problems that plague the current processes. The only way to gain such understanding is by having knowledgeable people to talk to.

What does this say for choosing the right scope? If the workforce is very inexperienced, there are two risks. First, it may take an inordinate amount of time to analyze the process because the expertise needed to get appropriate information may be hard to come by. For example, conducting intensive interview sessions with several experts at a job position is easier than getting the same information from two dozen individual interviews. Second, this area may be one to avoid altogether because either the inexperienced workers add no value, or the potential results may be based on too great a risk factor because of lack of knowledge of the work being done. The risk of re-creating the same problems or introducing new ones is greater when there is a lack of detailed information about the current processes.

Step 4: Move the Beginning and the End

After completing Steps 2 and 3, you have not finished the exercise. The information gathered during this exercise may suggest that the scope is proper (e.g., lots of hand-offs, few computer systems, etc.). The question then becomes, "Is this the best scope?" A simple way to find the best scope is to move the beginning or the end of the process that you have chosen to see how doing so changes the process level indicators previously gathered. Of course, the beginning and the end can be moved further apart to make the scope larger and moved closer together to reduce the scope. The goal is to find the best scope, given the amount of time, money, and resources that are available for the project. As suggested before, you are running the ideal scope smack up against reality. One rule of thumb is that the scope can be as large as the span of control of the executive sponsor, but this is a faulty assumption. It only frames the boundary of the possibilities; it says that given enough resources, the problem can be fixed. As the scope assessment categories point out, there is more to looking at a problem than employing the mere power of an executive. The point once again is that many factors can cause failure, and the lack of executive leadership is only one of them.

As we have mentioned, Connie Roberts's organization is responsible for the order fulfillment process—order receipt through firm ship date. Assume that the initial scope has been chosen (Figure 5-2). Now let us change the scope to include the sales and marketing organization (Figure 5-3). The reasoning here is that the orders do not simply fall out of the sky. Customers want to order things because the

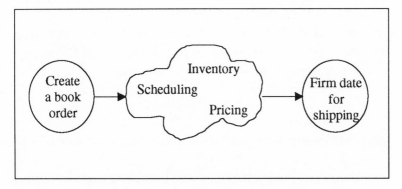

Figure 5-2. The first and last tasks in the initial book-ordering scope.

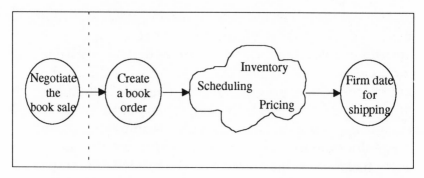

Figure 5-3. The new scope.

sales force was out hustling to make a sale. Maybe there are a bunch of redundancies between Connie's organization and the sales force. Or there may even be some efficiencies gained by tightly coupling the organizations. Whatever the reason is, let us assume the scope is moved.

Step 5: Iterate

After changing the scope by moving the process boundaries, Steps 1 through 3 need to be reiterated for the new delta. In the case of shrinking the scope, some information needs to be eliminated from consideration. Table 5-1 is an example of what may be seen after moving the scope on the above process. Keep in mind that there may need to be several iterations done before the right scope is achieved.

Step 6: Assess the Risk and Choose the Scope

The data in Table 5-1 make it clear that there is no scientific way to compare and evaluate risks. The approach suggested is a balancing act of what the specific constraints are on your project and how these categories either limit

Table 5-1. Results of varying the scope

Operational costs	10% growth per year	12% growth per year
Computer system costs	$3 million	$3.5 million
Hand-offs	24	N/A
Throughput	34 per month	N/A
Computer systems	5	5
People	300	350
Friends?	Yes	No
Culture	Good candidate	Bad candidate

or enable what needs to be achieved. In the easiest case, there may be a blatant difference in the potential scopes. For example, in Table 5-1 there is very little difference in many of the categories, but the assumption is that the people/management in the marketing and sales organizations will be unfriendly to the effort. This fact alone may sway you away from this particular scope.

NOW THAT YOU THINK YOU HAVE IT

Once you have chosen a scope, it is time to begin the next phase of the project, where a detailed assessment of the processes and underlying systems takes place. At this point, you should feel somewhat uneasy about the scope of the project. It is a healthy sign if you feel that you have bitten off more than you can chew. Before we discuss the products or outcomes generated through the "discover" phase of the project, a few words are needed about where this information will be helpful later in the project. First, this information is useful in providing a business case or other cost justification that may arise, as it has high-level indicators that management can relate to. Second, it is useful in quantifying why this scope was chosen over some others. Frequently, there is tremendous 20/20 hindsight, and this should curtail second-guessing. Finally, it can be used if the need arises to take another look at the chosen scope. Do not be afraid to reevaluate the situation if there are changes in the business as the project continues.

THINGS TO WATCH OUT FOR

Swollen Middle Management

One of the biggest impediments to a reengineering effort is an overstaffed middle management. Although the pure people numbers may impact the overall time required to implement a project, an overabundance of managers may make the implementation impossible. The problem is that these managers usually wield enough power to influence their immediate management. In addition, they assume that the problem is never within *their* shops—"If the people upstream in the process would only do what they are supposed to, we would not need ten people to redo their work." The best way to fight this problem is to bring it to the attention of the champion of the effort as well as the executive sponsors. Too often problems are talked about as if they manifested themselves only in the lower levels of organizations, when they are sometimes simply created and further supported by bloated management structures.

Unions

Unions create special challenges to reengineering efforts. The first impression that many union members have is that there is going to be tremendous downsizing. Hence, they feel the need to resist it. The reengineering team and its champion must be up front about what reengineering is as well as some of the possible outcomes. A natural result may be the downsizing or the restructuring of jobs, but it is best to let this be known. A key way to differentiate a reengineering effort from the changes that typically lead to downsizing is that the staff reductions in a reengineering project are driven by a well-thought-out business design. Instead of looking for reasons to automate or to downgrade jobs as a cost-saving measure, management should be asking the union workers for their input on process redesign.

Phase **II**

Hunt and Gather

The reengineering effort is now under way. You have assessed the damage, chosen a portion of the business to reengineer, and carefully selected a team to begin the project. Now what?

Well, now comes the fun part. This is the point in the project where you find out—really *uncover*—how your business actually runs. You should view the entire reengineering process as an education of sorts. You will learn things ranging from the social interactions found in small teams to the amazing complexity of developing and delivering large information systems. Consider the "hunt and gather" phase of the project as the grammar school of your reengineering education. As in grammar school, the main goal is for you to gain a basic understanding of how things work (including people). It will be exciting and challenging, but a long hard road will be ahead of you—and there will be no breaks for a snack and a nap.

In this phase, we will examine ways to understand the root causes of the current problems, as well as ways to analyze the business process you will be reengineering and the information systems supporting that process. Later phases will deal with the redesign and implementation.

In *Business Process Improvement,* H. J. Harrington writes:

> The biggest opportunity you have to improve the bottom line comes from improving your business processes. . . . In many companies, management can make more profits by cutting poor-quality cost in half than by doubling sales. This can be accomplished without hiring one new person, building one new building, or finding one new customer. There are dollar bills lying all around your organization. All you need to do is go out and pick them up and put them in the bank. (p. xii)

Although incremental benefits can be gained without additional cost, the same does not hold true for the quantum improvement of a reengineering project. Most reengineering projects require an investment in analysis of the existing environment, design of the new environment, and implementation of new information systems. However, Harrington's image of dollar bills on the shop-room floor is particularly vivid. As you proceed with a detailed analysis of the existing process and systems, you will see not just dollar bills but fifties and hundreds lying all about you.

During the detailed examination of the current business, you will likely uncover inefficiency after inefficiency. You will identify opportunities for reducing rework, administration, management layers, hardware, software maintenance, and on and on. However, there are three points to keep in mind:

1. The goal of the organization is to make money for the corporation.

2. Local optimizations may not produce the desired improvement in the extended business process.

3. You began the reengineering project to effect radical improvement, not to fix individual problems in the current environment.

As you proceed, you should constantly be questioning the value of a task in the business process. How does it relate to the goal of making money? Does the task exist because of a perceived rather than an actual need? Managers often create things to manage because they believe that it is part of their job. A common example is the proliferation of administrative reports. Systems operators sacrifice a veritable forest to produce reports that help managers do their job. You should be examining that job and its tasks as they relate to the goal.

Even if a task is necessary, that does not mean that it is a good candidate for improvement. Replacing the faucet and the sink will not affect clogged pipes and may, indeed, cause the sink to overflow even faster. The same holds true for process improvement. Eliyahu Goldratt and Jeff Cox describe this idea in *The Goal:* "We must not seek to optimize every resource in the system.... A system of local optimums is not an optimum system at all; it is a very inefficient system." In other words, get the big picture of how the whole process works and how it could work in the future before you focus on improving particular pieces.

The consequences of relating tasks to the goal and of having a global view of the process will be explored in detail in Chapters 7 and 8. First, Chapter 6 discusses how to get a handle on overall problems. How much are they costing the business? How easy are they to fix? Are they contained in a specific work group or confined to a small part of the process?

Much of this phase of the project is about striking a balance between analysis and action. You need to understand how the business works and why things are broken, but you also have to move on to the redesign phase so you can experience some benefits. It will not help your business if you have documented every aspect of the current environment but have lost your customers, your people, and your money. The next three chapters are aimed at teaching you how to separate the wheat from the chaff during the analysis phase, preparing you for the actual redesign.

WHY ALL THIS ANALYSIS?

Before describing some techniques for understanding the current process and its systems, we think it necessary to explain why a full three chapters of this book are dedicated to analyzing the current environment. Some reengineering literature[1] cautions against spending too much time documenting the current process and even recommends starting with "a clean sheet of paper."

This simply will not work for complex processes. For simple or well-understood processes, such as pizza delivery or accounts payable, you do not have to spend much time examining the current process. *At a minimum, however, you must know the root causes of the current problems and the value-added tasks in the process you are redesigning.* If you do not understand the root causes of the problems currently plaguing the business, you will be likely to experience the same problems with your redesign. If you do not recognize the work that is truly adding value, then you may not include that work (or its equivalent) in the new design. A clean sheet of paper only makes it easier to repeat past mistakes and fail to capitalize on past successes.

A cursory review (or no review at all) of the current environment leads people down the wrong path. In *The Fifth Discipline,* Peter Senge discusses a trap that many people fall into called the "leap of abstraction"—applying a broad generalization based on few facts. When this is done, conclusions often are incorrect. The team will be confronted with the opportunity to make these broad generalizations. The people they will rely on for information about the process and computer systems all will have a solution—"If we just did <X>, everything would work." It is imperative that the team go beyond a limited review of the current environment, and that they analyze to the point of understanding.

[1]For example, AT&T's *Reengineering Handbook* spends less than one of its 146 pages on understanding the current environment.

6

Studying the Problem

After the team has been formed and a process has been selected, the first job of the team members is to make sure they are focusing on the right problem. They will need to get a handle on how bad things are and how their process compares to similar processes in other companies. The work described in this chapter lays the groundwork for a vision of how the business could operate in the future.

Before conducting an in-depth analysis, the team first must get an overall sense of the business process and its problems. The methods include a variation of work sampling used to identify bottlenecks quickly and to gain a rough approximation of the cost of individual tasks. Also, this chapter describes the use of benchmarking for getting comparative data on the process and gauging your position in the marketplace.

This quick analysis of the situation serves to verify that you have chosen an appropriate scope and then points out what to do next. Be prepared for a change in scope based on your early findings. This should not be necessary if all went well in applying Chapters 4 and 5, but the methods described in this chapter go beyond the categories in those earlier chapters and provide more quantitative data. In a typical functionally organized business, the original charter may be to find out what is wrong with system X (or process X, or team X). Your early results will help you to verify whether the problems are cross-functional and whether the scope you selected in Chapter 5 is the right one.

We recommend that this up-front analysis last only a month or so. The intent is not to research every problem but to get a solid overall sense of how things work and how they might be better. The reengineers on the team will be doing most of this work, but the public relations people and the champion are "greasing the skids" by making it clear why a group of people are poking around in the business operations and why they should not be hindered.

WALKING THROUGH THE PROCESS

To gain even a high-level understanding of the current process, you must talk with the people who execute the process—the "operations" people. No matter how many managers or vice-presidents or whatever are involved in the business, the bottom line is the work performed by operations personnel. Dr. Deming inspired us, in part, by his approach of collecting information about a process from the people who were actually executing the process.[1]

---

[1] *Out of the Crisis* is chock full of quotes from operations personnel—factory workers, production workers, plant managers, etc.

They are the people who talk to the customer, who work with the raw materials from suppliers, and who interact with the computers. They are your most important source of data.

Understand that at this point in the reengineering project you are interested in an approximation of the process. You will be applying more detailed, scientific methods in the next chapter. At this stage, however, consider your work sampling to be an informal survey of the process and its support. This survey consists of three steps: collecting current documentation on the process, observing and interviewing operations personnel, and documenting the steps in the process and their costs. After walking through the process for a few weeks, you should have a first cut at the tasks in the process and how they are linked, how long each task takes, and a list of problems and potential improvements. Again, this is not a particularly structured list, but it will help you focus your attention as you perform the detailed analysis described in the next two chapters.

In his seminal article "Reengineering Work: Don't Automate, Obliterate," Michael Hammer summarized his view of analysis of the current process:

> The team must analyze and scrutinize the existing process until it really understands what the process is trying to accomplish. The point is not to learn what happens to form 73B in its peregrinations through the company but to understand the purpose of having form 73B in the first place. . . . Rather than looking for opportunities to improve the current process, the team should determine which of its steps really add value and search for new ways to achieve the result.

The goal is to analyze the current environment to the point of understanding the root causes of the current problems so as to avoid them in the new business design. Ignoring the way work is currently performed, starting with a "clean sheet of paper" or focusing on changing corporate strategies, "bespeaks a profound contempt for the daily operations of business. . . . If American companies want to become winners again, they will have to look to how they get their work done. It is as simple and formidable as that."[2]

Collecting Current Documentation

Most processes have some written material describing the process steps. Most of this material is incorrect or at least incomplete. Every set of process documentation that we have seen has been annotated *ad absurdum* by the operations staff. Despite the efforts of personnel dedicated to keeping the operating and training material up-to-date, most of a company's valuable experience is penciled in the margins. Nevertheless, the existing documentation is a useful guide during your initial walk-through of the process. After a few days, you will quickly learn how closely the documentation tracks to reality.

Working with Operations Personnel

Over the course of two to four months in the "hunt and gather" phase, you will be spending a great deal of time with operations personnel, working to understand how they do their jobs. In the first few weeks you get an overall sense of how the common scenarios are handled by observing the staff and asking questions.

---

[2]Hammer, Michael and James Champy, *Reengineering the Corporation*, pp. 25–26.

This is not a particularly structured observation. Do not slip on a white coat and a stopwatch and play "observer." If you do, the people will feel like laboratory rats, and you will not get the information you need. Instead, treat the people executing the process with respect. After all, they are the ones responsible for actually providing the product or the service that the company is based on. Let them know why you are interrupting them and how you will use the information they give you. As you watch them work, ask questions about why they perform each task and how their job relates to that of other people in the process. If you notice a problem, ask them what they think its cause is, and exchange ideas with them on ways to improve things. Most people like to talk about their jobs if someone shows genuine interest in them, and their input and cooperation are essential to the success of the project.

You should begin observing the inputs to the process and follow the sequence of tasks and people until the outputs are produced. (At the same time, you can work from the end backward.) As you follow the course of the process, you will identify distinct tasks. Working with the operations person who executes a given task, establish a name for that task, and have the person describe how long it takes to execute the task. Because individual tasks probably will not take the same amount of time every time, have the individual operators assign a minimum, maximum, and average amount of time it takes to perform each of their tasks.

As you proceed, you will be collecting tasks and execution times. Pay particular attention to process inputs/outputs, places where organizational boundaries are crossed, and hand-offs within the process. These are the places in the process where most errors are introduced; so you should try to understand these areas best of all.

Documenting the Process

As you work with operations personnel you should capture your understanding of the process with simple flow
charts and brief textual descriptions of each task. Review
these documents with the operations staff as you proceed.
Be prepared for some frustration, however, especially
with complex processes such as provisioning telecommunications equipment. You are likely to make translation
errors, or it may turn out that your picture captures only
the 1 percent case that someone has described. It often
takes multiple iterations to capture the work accurately.
(Cross-checking results with a number of workers is essential in these cases. Often, the senior, experienced staff—
such as the insiders on the reengineering team—can clarify
complex issues.)

For each job function involved in the process, find out
the cost per hour (e.g., by dividing the hourly staff costs
supporting that function by the number of task executions
per hour). Then, multiply this cost by the task times you
collected. This will give you an approximate, and most
likely a conservative, estimate of the staff-cost of each task.

As you document this information, certain tasks will
jump off of the page, screaming to be reengineered out of
the process. There are plenty of candidates:

- Review tasks ("I review all of Bill's work to make
  sure it's correct before I proceed").
- Re-keying tasks ("I take these three reports and type
  ten fields into this screen").
- Courier tasks ("Now I take the folder to the guard so
  he can bring it to the other building on his rounds").

The documentation will be used to create the initial
process model described in the next chapter. The ad hoc
lists of problems, improvement ideas, and screaming tasks

will help you focus your attention on certain areas when you do the detailed analysis and will begin to trigger ideas for the redesign phase described in Phase III.

BENCHMARKING

In walking through the process you have either verified that you have chosen the correct scope, or you have gone back to Chapter 5 to see why your original choice missed the mark. Benchmarking is a way to compare your business process with those of other companies and to get ideas on how to improve or redesign your process. In a reengineering project, you will look for two outputs of a benchmarking study: characteristics of best-in-class companies and direct measurements of their processes. The study can be performed by your company or by consultants *while* you are walking through the process. You will use the study in the redesign described in the "innovate and build" phase.

A good definition of benchmarking is provided by Robert Camp in his book *Benchmarking:*

> Benchmarking is the search for those best practices
> that will lead to the superior performance of a
> company . . . and [which] allows a manager to compare
> his or her function's performance to the performance
> of the same function in other companies.

Why should companies with "best practices" give you any information? These companies, like your own, are interested in getting even better. By exchanging information with companies in other industries they can learn about new ways to solve problems that might apply to their own business. Robert Camp gives the example of Japanese typewriter manufacturers visiting grocery outlets and incorporating barcode technology into their assembly lines. The best-in-class companies understand that many business problems are the same regardless of the industry.

Table 6-1 summarizes attributes of the different kind of benchmarking. Reengineering projects tend to rely on innovation, and so the focus is on best-in-class companies ("industry leaders") and on companies in other industries ("generic processes").

With some preparation and negotiation, you can usually set up surveys or even site visits of best-in-class companies to get information on how those companies solve problems and succeed. Usually, you can also get this information from your direct competition. Although your competitors will not give you information about their processes, that information is often available to industry consultants (for a sizable fee, of course).

The primary aim of your benchmarking study is to determine the characteristics of best-in-class companies

Table 6-1. Key benchmarking characteristics*

Benchmarking operation	Relevance	Data collection ease	Innovative practices
Internal operations	√	√	
Direct product competitors	√		
Industry leaders		√	√
Generic processes		√	√

*This table shows the different sources of benchmarking data, whether they are directly relevant to your business, how easy it is to collect the data, and whether the data are a good source of innovative ideas. (Adapted from *Benchmarking* by Robert C. Camp.)

that have similar processes. For example, Robert Camp describes the best practices for companies with warehousing processes, including shipping, receiving, storing, and other subprocesses. The best of those companies shared innovative practices that utilized technology to minimize errors and rework. Two examples of such sharing are the receiving and shipping functions.

The receiving docks used on-line access to inventory and purchase order data, enabling the receiving personnel to reconcile deliveries to outstanding purchase orders, update the inventory, and update the status of the purchase order *all in a single system with one source of data entry.* (This eliminated receiving items for which there was no purchase order and, more important, eliminated the need for an inventory data entry group and a purchase order update group and a reconciliation group.)

Shipping used automation both to prepare the shipping documentation and to sort the packages to the correct carriers. This reduced the amount of time and effort needed to perform these tasks and eliminated the errors due to human transcription of delivery information. Again, the elimination of errors eliminated the need for dedicated rework/error correction groups.

These practices apply to almost any company with warehousing needs, ranging from L.L. Bean sweaters to parts for Xerox copiers. You will use the best-in-class characteristics from your benchmarking study as an input to the redesign phase described in the "innovate and build" phase.

The secondary aim of the benchmarking study is to collect data on the processes of your competitors. These data usually include unit cost, customer satisfaction, and delivery intervals, as well as measurements that may be specific to your industry. This information will serve as one of the inputs to your vision of how the business could operate. Keep in mind that there is no simple algorithm that translates the measurements of your competition to a new set of metrics for the reengineered process. Devising measurements of the new process is discussed in Chapter 9.

THINGS TO WATCH OUT FOR

Problems ''Getting Behind the Lines''

As we said above, you will need the champion to make it clear to the operations management that your reengineering activities are for the good of the company. Operations groups are naturally threatened when outsiders are brought in to streamline their operations. The result is usually a critical report of their work and a downsizing of their staff. The champion and the team itself must work at building trust between themselves and the people who may be affected by their work. You need to solve problems of gaining access to information in the first week, or your effort will suffer. (This is another prime reason for having insiders on the reengineering team.)

Overanalysis

During this step, your aim is not to fix the myriad problems with the current environment or to find out how the operations personnel handle those once-in-a-lifetime scenarios. The purpose of your foray into the business operations and your study of other companies is to gain a gestalt sense of what is wrong and how things could be better. Spending more than four weeks in this activity or gathering reams and reams of documentation for detailed study is a sign of overanalysis.

Insufficient Resources

When you are studying the problems, there are so many activities—interviewing, observing, reading, visiting other companies, documenting findings—that the small reengineering team may be overwhelmed. Try as much as possible to divide and conquer. Send one or two individuals to carry out specific tasks, and then convene as a team to compare notes. For certain tasks, such as creating the benchmarking study, the use of consultants often is worthwhile. Good benchmarking consultants have better access to other companies than your team might have, and they are experienced at creating these studies. Outsourcing tasks like these (and getting better-quality results) is a good way of reducing the workload of the reengineering team.

7

Understanding the Current Process

A solid understanding of the facts is as important as executive commitment and a grand vision of the future. You are reengineering because you need to do things very differently from the way you are currently doing them. If you cannot identify the root causes of the current situation, you will have little chance of making changes for the better.

This chapter is about identifying what the process *should* be doing and the reasons *why* the process is not doing those things. To achieve these objectives, you will need the following:

- A means of working closely with customers.
- A solid understanding of variation in a process.
- Statistical tools for analyzing process data.
- Software to simulate the execution of the process.

Many of the techniques described in this chapter have been applied successfully in companies involved in Total Quality Management (TQM) programs.[1] A relationship exists between reengineering projects and TQM methods because many current value-added tasks may well remain in the new design. The TQM methods help you identify and measure the value-added tasks. Process breakthroughs are discovered by using methods described in later chapters. These techniques focus on process and organizational boundaries, consolidation of activities, and introducing enabling technology.

Knowledge of process variation and statistical tools is critical in helping you distinguish symptoms from root causes. The process simulation will help you understand the dynamics of the process and will be useful in comparing design alternatives in the "innovate and build" phase of the project. Simulation has long been used to model manufacturing processes, but the concepts and principles apply equally well to service processes. Simulating many executions of a process highlights the bottlenecks, costs, utilization of people's time, and throughput.

To understand the current process and the current information architecture (described in Chapter 8), the reengineers on your team will be working closely with customers and operations personnel for two to three months. The exact amount of time depends on the complexity of the process. You will extend the list of root causes to be addressed in the redesign phase, and you will use the process documentation described in Chapter 6 to create

---

[1]The relationship between TQM and reengineering is explored in an article by Thomas H. Davenport entitled "Need Radical Innovation and Continuous Improvement? Integrate Process Reengineering and TQM," in the May/June 1993 issue of *Planning Review.* The July/August 1993 issue of *Planning Review* describes how some companies involved in reengineering projects, such as Marrion Merrel Dow (the prescription pharmaceutical business unit of the Dow Chemical Co.), view business reengineering as an extension of their company's TQM program.

the initial mechanized model of the process. The model will be used in the redesign phase to quantify the effects of different changes in the process, technology, and staffing.

TALKING WITH CUSTOMERS

To understand what a process should be doing, there often is no better approach than talking with the people who use the products of that process. The statistics and process modeling described in this chapter are excellent tools for understanding why a process behaves the way it does, but they do not shed light on what the process should do.

Customers are not interested in your organizational problems or your poor systems architecture or your problems with the mail room. They have limited interaction with your processes, and they only care about getting value for their money in terms of products or services. Customer satisfaction surveys and direct interviews with customers both yield useful information on the expectations and desires of the people who are paying for the products of your process. Just as you used benchmarking to identify your position in the marketplace, you should consider a customer satisfaction survey to identify your position with the customer. (You may be able to use existing data, or you can commission a new survey to get the data.) Such a survey usually will give you information on what customers want ("I usually need the supplies *that minute* or at least that same day"), the key weaknesses of the process ("It takes too long"; "No one seems to know where my order is"), and some opportunities for improvement ("I wish I could enter my order by telephone"). Without any need for detailed analysis, this is a good way of getting unfiltered feedback on the performance of your process.

Direct interviews with customers provide much of the information you will find in surveys, but the interactive nature of the interview allows you to ask questions ("Why do you need to call for status?"; "What information do you use to make decisions on what or how much to order?"). You may only have the resources to interview a small sample of customers, but the combination of a broad survey and detailed interviews will give you an appreciation of what customers expect and want from your process.

In *The Customer-Driven Company*, Richard C. Whiteley advocates that you "saturate your company with the voice of the customer" through a variety of mechanisms. In addition to surveys and direct interviews, for example, he proposes that companies solicit and analyze customer complaints:

> A study for Travelers Insurance showed that persuading people to complain could be, in fact, the best business move a company could make. Only 9 percent of the noncomplainers with a gripe involving $100 or more would buy from the company again. On the other hand, when people *did* complain and their problems were resolved quickly, an impressive 82 percent *would* buy again."[2]

In addition, Whiteley found that companies often uncovered ideas for new products or gained valuable insight into business problems as a result of customer complaints.

Other successful methods for listening to the voice of the customer include focus groups, customer councils,

---

[2]Whiteley, Richard C., *The Customer-Driven Company*, p. 40. The results of the study were published in *Travelers Tribune*, February 1989.

and even videotaped sessions of customers actually using a company's products or services. Few methods are as powerful as watching a customer complain, or deal with an inferior product, or fumble with awkward packaging and instructions. "It's like the ghost of Christmas Present showing Bob Cratchit's house to Ebenezer Scrooge. Nothing else can change an executive's [or employee's] attitude so dramatically."[3]

VARIATION IN A PROCESS

Once you have a grasp on what value the process should provide to customers as well as a sense of why it is not providing that value, you can look for specific reasons for your process's poor performance. "Normal" fluctuations of process measurements lead many a manager on a wild goose chase. In *Out of the Crisis,* Deming rails against management's ignorance of the causes of variation and points to it as a central problem in management and leadership: "A fault in the interpretation of observations, seen everywhere, is to suppose that every event (defect, mistake, accident) is attributable to someone (usually the nearest one at hand) or is related to some special event."

In explaining the causes of variations, Deming distinguishes between common causes and special causes. Faults of the system are common causes. Variations due to common causes are the result of the process itself and cannot be eliminated unless the process is modified. Faults due to serendipitous events or to individual performance are special causes. He estimates that 94 percent of all troubles

---

[3]Whiteley, Richard C., *The Customer-Driven Company,* p. 52.

belong to the system, and only 6 percent are due to special causes. The following is a priceless exchange from *Out of the Crisis:*

> "Bill," I asked of the manager of a company engaged in motor freight, "how much of this trouble [shortage and damage] is the fault of the drivers?" His reply, "All of it," was a guarantee that this level of loss will continue until he learns that the main causes of trouble belong to the system, which is for Bill to work on.

We have found, as Dr. Deming described, that managers tend to react to fluctuations in process measurements by "adjusting" the process. However, this tends to cause greater variability and frustration for workers and managers. The question remains, however, how does one know when the system (and not an individual or a special event) is to blame? The answer is found in statistics: processes are not deterministic, but are composed of probabilities and distributions.

This is best illustrated by a simple example adapted from *Out of the Crisis.* Hold a funnel over a flat surface and drop fifty marbles through the funnel, one at a time. Mark where each one lands. You will have fifty marks showing the distribution or pattern of marble falls. The results of one trial are shown in Figure 7-1. (*X*'s mark the spots where the marbles landed.)

Try the experiment again. This time, however, whenever a marble does not land directly below the funnel, move the funnel over the point where the marble landed. Again, mark the spot where each marble lands (using *O*'s this time). Figure 7-2 shows the results of this trial overlaid with the original results. As Deming notes, "The successive drops of the marble resemble a drunken man, trying to reach home, who falls after each step and has no idea which way is north."

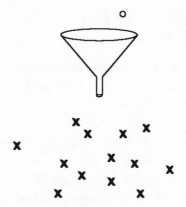

Figure 7-1. Results of the funnel experiment.

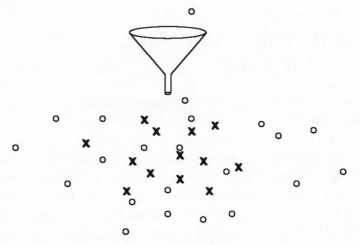

Figure 7-2. Disastrous effects of moving the funnel.

The funnel experiment illustrates four points:

1. There is variation from event to event (the spot where the marble lands varies each time it passes through the funnel).

2. You cannot predict individual events (you really have no idea where the next marble will land).

3. Under stable conditions (such as holding the funnel steady), a pattern emerges.

4. Tweaking the process (even in the hopes of making it better) can increase the variation and could have disastrous effects.

Thus, variations in processes are to be expected, but as you collect data in a stable process, you will see a pattern or distribution of events and should be able to sort out the special causes from the common causes. You can do this by using some of the simple statistical tools described in the next section.

STATISTICAL TOOLS AND ROOT CAUSE ANALYSIS

Statistics have been applied successfully to process problems for a long time. More than seventy years ago, Western Electric was seeking to improve the quality of its products. As the manufacturing arm of the Bell Telephone System, Western Electric produced what was then sophisticated telecommunications equipment; but variations in the product were causing problems for the company. In 1924, Dr. Walter Shewhart ushered in a new era of quality improvement by developing the ideas of statistical process control.[4] There are volumes of material on how statistical process control can be used to improve quality, but this section will focus on just one of Shewhart's ideas—the control chart—and how it can be used to identify root causes of some process problems. This is meant to give

---

[4]These ideas were later published in *Economic Control of Manufactured Product,* D. Van Nostrand, 1931.

you a sense of the power of these kinds of tools. Other tools, and ways that they can be used in root cause analysis, are described briefly.

The Control Chart

A process is "in control" if common causes only are operating, and no special causes are influencing the variations. The control chart helps you separate the noise (variations from event to event) from the fluctuations that arise from special circumstances.

As an example, consider a fast-food drive-through process. In such a process, special causes are usually the *differences* between workers or equipment. Leaving a new, poorly trained employee to handle the lunch-time rush would be a special cause. Three of the four fryers breaking down would be a special cause. Once the root causes of these problems are discovered, they can be addressed by dealing with the special circumstance. Common causes relate to the process itself. If all workers have to walk to the basement to get supplies and must leave the counter unstaffed, that is a problem common to the system. There are no special circumstances in this case; the process actually is designed to have workers leave the counter to get supplies. To address these kinds of problems, the process must be changed.

The control chart depicts the mean performance of a process over time. In Figure 7-3, the mean service time (time to service a customer) is plotted over a course of two weeks at half-hour intervals. The "controls" (upper and lower control limits) are shown by drawing lines that are three standard deviations from the overall mean service time. In other words, take all of the service times over the two weeks and compute the average; this is the centerline.

Then, calculate the standard deviation, and simply draw a horizontal line that is three standard deviations from the centerline.

The three graphs in Figure 7-3 show three things to look for in a control chart: trends, cycles, and freaks. Figure 7-3a shows an upward trend in service times. Trends indicate a slow shift of the process. In these cases, look at your other process measurements for parallels with the trend. Cycles are unnatural patterns that occur periodically. In Figure 7-3b, the chart was out of control every Saturday during lunchtime. Freaks, shown in Figure 7-3c, are individual cases—there is no pattern to the out-of-control measure-

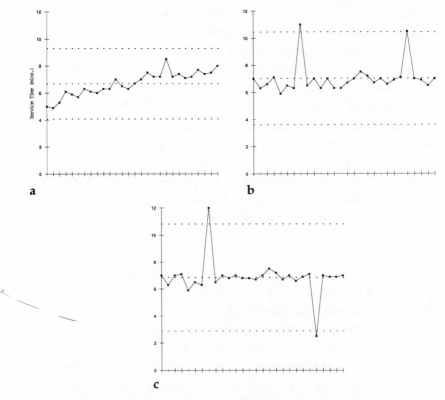

Figure 7-3 (a–c). Typical control charts showing trends, cycles, and freaks.

ments. These data usually are due to truly special circumstances, such as a sick employee.

Using a control chart is one way of determining whether you can eliminate a problem by redesigning the suspect portion of the process, or whether the problem is due to some element that is out of your control. Being able to make this kind of determination prior to the "innovate and build" phase will help you avoid making unnecessary process changes.

Other Tools

There are seven tools, often associated with Kaoru Ishikawa,[5] that are commonly used in root cause analysis. Despite their simplicity, these tools (one of which is the control chart) are effective ways of organizing data to reveal the true source of a problem. Each brief description of the other six tools is accompanied by a sample illustration.

Cause-and-Effect Diagram
Also called a fishbone diagram, this is useful for organizing potential causes of a problem or the elements needed to reach a goal.

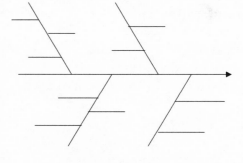

---

[5]These tools and others are explained in *What Is Total Quality Control?* and *Guide to Quality Control,* both by Kaoru Ishikawa. An excellent, practical description of these tools is also included in *The Deming Management Method* by Mary Walton.

Flow Chart
This chart is a simple graphic way to illustrate the logic in a process.

Pareto Chart
This chart is useful in determining priorities as you group your data into categories. Pareto charts typically show how a small number of problems are responsible for a large percentage of the total defects or costs.

Run Chart
This chart shows data over time to help identify time-dependent trends. Note that some variations observed in a run chart may be due to special causes.

Histogram
This is used to measure the frequency of occurrences of some event. These charts are helpful in determining whether a process is in control.

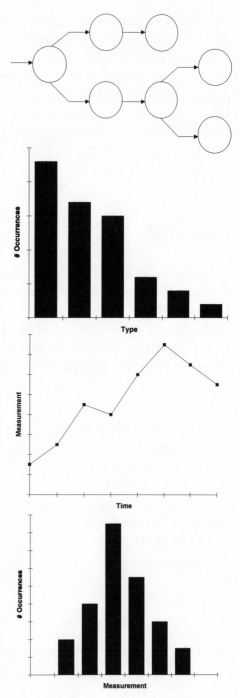

Scatter Diagram

This diagram illustrates the effect of one variable when another variable is changed. These graphs are useful in testing for a cause-and-effect relationship.

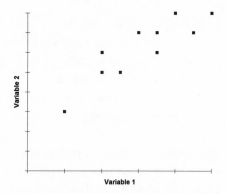

Some Applications of Root Cause Analysis

At this point, you might be asking, "If I am looking for radical breakthroughs in the process, why am I wasting time with process control?" We reiterate the point made in the introduction to Phase II: If you do not understand the reasons why you are in the current situation, you are likely to repeat the mistakes and problems that caused you to reengineer the process. Moreover, if you do not understand how to measure a process and improve it incrementally, your redesign eventually will spiral out of control, bringing you back to where you started.

Root cause analysis begins with data collection and analysis (such as that done with control charts). However, finding the real root cause of a problem you have identified using a control chart is not a science; there is no algorithm other than repeatedly asking why things are the way they are. Imagine that a problem in your drive-through process is mistakes in orders. People are getting large fries instead of small fries, cheeseburgers instead of bacon cheeseburgers. Your data show you that the process is consistently producing wrong orders. Imagine trying to determine the root cause:

- *Why?* "Because the customers don't specify exactly what they want."

- *Why don't you ask them?* "I used to ask them, but it would take too long and they'd get mad."
- *Why?* "Because they would have to repeat their answer several times."
- *Why?* "Because I couldn't always make out what they were saying."

The root cause then is not laziness, not bad procedures, not mumbling customers—it is just a faulty intercom system. Tweaking the drive-through process instead of analyzing the data would only have produced more variations and more incorrect orders.

Below are some examples of root causes we came across on a recent project:

- Functional organizations instead of process-oriented organizations.
- Disjointed work groups linked only by phone or fax.
- Insufficient or inadequate training.
- Complex customer interfaces, such as lengthy forms.
- A fragile information systems architecture (which makes it difficult to get support for new processes and tasks).
- Unpredictable arrival rates.

MODELING THE PROCESS

Even in a simple process, statistics are at work all the time, with probabilities and distributions guiding what actually happens.[6] Modeling the process helps you under-

---

[6]The terms stochastic systems and queuing theory often are used in describing the statistical aspects of a process. A stochastic process is one in which the variables describing the process take on values according to some probability distribution. The mathematics behind stochastic systems is called "queuing theory."

stand the process dynamics—how the statistics affect the actual execution of the process. Modeling can be defined as "the process of developing an internal representation and set of transformation rules which can be used to predict the behavior and relationships between the set of entities composing the system."[7] In other words, if you know the tasks in the process and how they relate to each other (e.g., distribution of task execution times), you can predict how the process will behave in different situations.

Consider the simple process of the drive-through at a fast-food emporium, depicted in Figure 7-4. Such a process can be described with four pieces of information:

1. The arrivals of customers. Customers arrive at the low-fidelity speaker in a probabilistic manner. That is, with some fluctuation, people arrive in a fairly predictable pattern. There are more people at mealtimes, for example, and few at 3 A.M.

2. The time it takes to fill each order. It does not take much time to fill an order for large fries and a chocolate shake.

3. The number of people working the counter.

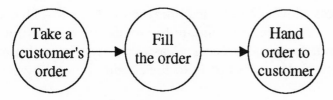

Figure 7-4. The drive-through process.

[7]Franta, W. R. (1977), *The Process View of Simulation*, Elsevier-North Holland, Inc.

4. The customer-selection method. At this particular fast-food restaurant, the method is first-come, first-served. It could have separate lines for large orders, however, or for special customers.

Now, merely eyeballing this process will not allow you to predict anything. How many people should staff the drive-through? How long will the lines get during lunchtime if I only have one person? What would happen if I set up a "special orders" line? Most managers use "experience" as their guide. We have found, however, that the key to answering these kinds of questions is process simulation.

Simulation is a means of building the process model in software. The computer allows you to simulate the fluctuations of the process, to "execute" the on-line process many times with different sets of inputs, and to model the actual capacity of individuals (by including interrupts, off-time, etc.). If you knew the information described above, you could develop a simulation of the drive-through process in a day or two.

Models are meant for even more complicated processes. There have been models for policy responses to AIDS[8] and even for life itself.[9] A number of books tell you how to build one.[10] At this point in the reengineering project, you can use the process documentation described in Chapter 6 to create the initial model. (You have already defined a process flow and assigned time distributions to each task.) Then, you need to validate the model by comparing its output to actual process measurements. You will iteratively make refinements to the model, by adding more

---

[8]Whicker, M. L. and Lee Sigelman (1991), *Computer Simulation Applications*, Sage Publications.

[9]SimLife, a software package developed by Maxis, allows the user to change certain aspects of the environment and watch the effects on life on the planet.

[10]For example, see Shannon, R. E. (1975), *System Simulation: The Art and Science*, Prentice-Hall.

detail or altering times, until your model approximates the current process.

Figure 7-5 summarizes the inputs and outputs of your simulation. In this chapter, the bottlenecks and the resource utilization are key. For processes that are slow or costly, these items will point you to root causes. The other outputs—throughput, costs, and so on—are measurements you will try to change as you experiment in the redesign phase and simulate different process changes.

Goldrath and Cox's *The Goal* contains a section (pp. 148–160) that succinctly describes the need to identify and deal with bottlenecks in your process—resources that are overutilized. In this section, a consultant (Jonah) is explaining that the bottlenecks throttle throughput and so affect the entire system. Time spent at a bottleneck is the most valuable time in the process.

"What you have learned is that the capacity of the plant is equal to the capacity of its bottlenecks," says Jonah. "Whatever the bottlenecks produce in an hour is the equivalent of what the plant produces in an hour. So . . . an hour lost at a bottleneck is an hour lost for the entire system." . . . "Tell me something," asks Jonah. "How much does it cost you to operate your plant each month?"

Lou says, "Our total operating expense is around $1.6 million per month."

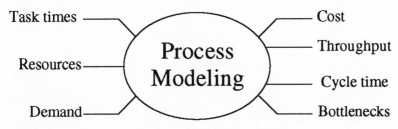

Figure 7-5. Process simulation inputs and outputs.

"And let's just take the X machine [a bottleneck] as an example," he [Jonah] says. "How many hours a month did you say it's available for production?"

"About 585," says Ralph.

"The actual cost of a bottleneck is the total expense of the system divided by the number of hours the bottleneck produces," says Jonah. "What does this make it?"

Lou takes out his calculator from his coat pocket and punches in the numbers. "That's $2,735," says Lou. "Now wait a minute. Is that right?"

"Yes, it's right," says Jonah. "If your bottlenecks are not working, you haven't just lost $32 or $21 [the cost of an individual machine for an hour]. The true cost is the cost of an hour of the entire system. And that's twenty-seven *hundred* dollars."

Lou is flabbergasted.

By showing you the dynamics of the process, the simulation model helps you identify the bottlenecks as well as resources that are underutilized. At this point you should identify the root causes of these resource utilization problems. You will read about how to avoid resource utilization problems in the "innovate and build" phase.

8

Understanding the Current Information Architecture

"Information architecture" is a lofty term for something very simple: the computer support of your business process. This encompasses all of the people involved in the execution of the process and how they use computers (ranging from mainframe applications to spreadsheets kept on a PC) to help them with their jobs. The word "architecture" is used to denote the structure and support that computer systems give to the information needs of a business process. However, remember that the Tower of Babel also had an architect.

The information systems problems of Instron Corp., a manufacturer of materials-testing equipment, are typical of almost all companies (and parallel our findings in the telecommunications industry):[1]

- High MIS operations costs.
- Long backlog of programming requests.
- Difficulty in making system modifications.
- High degree of user reliance on MIS staff.
- Insufficient integration of software modules.
- Inadequate timeliness of information due to batch transaction processing.
- Lack of commonality between information systems.

Business processes in need of reengineering typically are supported by a brittle information architecture. As with Instron, there may be a proliferation of independent, incompatible information systems. Worse, the systems may even have similar functions but are designed to meet each organization's own functional need and not the true business need. Tracking systems often are a good example of this. Instead of tracking the widget from receipt of the order through billing (in an order fulfillment process, let us say), the sales group will track groups of orders and order acknowledgments for big customers; the order-filling group will track individual orders; the warehouse people will track things at the item level; and the billing group will track the bills sent and complaints about bills. Then, despite all of this administration and CPU cycles devoted to tracking, no one will really know the status of the order. No wonder that the enormous expenditure on information technology has as yet led to little, if any, productivity increases.

Before you can radically improve the information architecture, you should understand what the individual sys-

---

[1]Loeb, Jeff, Highly configured products benefit from 'expert' order entry, *Manufacturing Systems*, July 1993.

tems do and how they do it: hardware and software architecture, interfaces, ease of accommodating new development, and so on. Most important, you must get a handle on the cost of these systems. Consolidating those tracking systems, for example, would slash system maintenance costs while improving the quality of status information and freeing development resources for other needs. However, if one of those tracking systems is the only way to tie in to corporate accounting, and it would be prohibitively expensive to redesign it, then you should be aware of that constraint as early as possible.

The goal of this chapter is to help you understand how computer systems are supporting the information needs of your process, and to pinpoint areas that need to be "rearchitected." Looking at the hardware and the software will give you only part of the picture. You must also gain an understanding of who is using these systems and what these people are doing—what value is added by that system function.

Within one to two months of studying the problems (described in Chapter 6), you should be able to pinpoint the weaknesses in your information architecture. Moreover, you should be able to quantify the cost of those weaknesses, ranging from the high maintenance cost of mainframe hardware platforms to the high staffing costs incurred when systems do not talk to each other. Together with the quantified understanding of the weaknesses in the current process, you will be equipped for the "innovate and build" phase.

HOW TO BEGIN

The people executing and managing the process often have little idea of how the computer systems work together to support the process. Typically, "the system" is viewed as a monolith, and people's knowledge of it goes only as far as their terminal. The reengineering team must dispel this monolithic image and discover the entire variety of existing computer support and learn exactly how people are using that support.

In this chapter, you will take two approaches to gaining this knowledge. First, you will follow the process flow and examine the use of computer support for each task. Using the methods for examining the process that were described in Chapter 7, you will construct a picture of the information architecture as you construct a model of the process. Seeing how each system is used will give you the lay of the land—a general understanding of which systems are critical and which are not, how well the systems exchange data, and how well the process is supported by technology. Further, this chapter will aid you in identifying the root causes of some common problems with information architectures.

Next, you will examine the interfaces, the hardware and software architectures, and the operating expenses of each system in order to quantify the information architecture costs and to determine the value of the systems to the corporation. You will use this information, along with the root causes that you identified in going through the process, in the redesign phase.

THINGS TO LOOK FOR AS YOU FOLLOW THE PROCESS

As you go through the process, observe how operations personnel are interacting with a system or with output generated by a system. Take careful note of four areas:

1. *Input:* What information is input, and how is the input accomplished?

2. *Output:* What information is generated by the system, and how do people manipulate that output?

3. *Exchange:* How do individuals or organizations exchange information with each other?

4. *Users/usage:* How many people in the different functions use the system, and how much is it tied to their job?

The next four subsections list common problems we have found in each of these areas. Some of these problems are mapped to root causes that we have identified in past projects, whereas others are intended to give you ideas for the new business design (e.g., things to avoid). The lists below are not complete, and you should apply the root cause analysis techniques described in Chapter 7 to verify the real causes of the problems you identify.

Input Problems

The Deming management philosophy emphasizes the need to work with suppliers to improve the quality of their products that serve as inputs to manufacturing processes.[2] The same principle holds true in any process, except that in most service processes the suppliers are producing data. In our experience, it is in the inputs to a process that most errors are introduced (and this is where there is an opportunity for breakthroughs). As you go through the process, examine each input critically. Can the inputs be generated mechanically? Has someone already input these data somewhere? How are you storing and

---

[2]Mary Walton's *Deming Management at Work* gives several excellent examples of companies working with suppliers to improve the input to the process, including a metallurgical company training its suppliers in statistical process control.

communicating the data? The following paragraphs list the most common problems we have seen with data inputs.

Manual entry of long or complicated sets of data; manual calculations: Long order forms, codes, and repetitive entry result in significant errors. Consider a typical purchase order with ten items on it. If each item is identified by a six-character code, and a data entry clerk has a 99 percent accuracy rate per character, the chances that the clerk will accurately enter the entire list of item codes are less than 55 percent.[3] These kinds of errors are prime sources of rework later in the process. Attempt to slim down the forms by searching for alternate sources of the data. Also, technologies such as on-line customer entry of data and optical character recognition can mechanize the transfer of information from the customer, eliminating the expense and errors associated with humans keying in data.

Re-keying the same information into multiple systems: Data redundancy across systems and insufficient system-to-system interfaces force humans to exchange data from one system to another by re-keying the data. If the data are used, say, for both inventory and billing (and so are expected to match up), this redundant entry can be a major source of errors. Without sophisticated distributed database techniques, the synchronization of redundant data across systems is an impossible task. Manual keying into multiple systems is clearly error-prone (as evidenced by the simple calculation above), and building interfaces between the systems is expensive and lends itself to systems' "rejecting" data from other systems.

Manual entry of status: This is a sign that the systems are not linked to the process. If people must type in what they did after they have done it (and there is more work to

---

[3]For those interested in the math, the chance of entering each character correctly is .99. Since there are ten items, each having a code of six characters, there are 60 characters, and $(.99)^{60} = .547$.

do), you can be sure the status information will be incorrect. See Chapter 12 for ways to link system functions with process tasks in order to track work-in-progress automatically.

Keeping paper logs: The proliferation of paper is a sign of a bad process and/or a bad work environment. Keeping historical logs that track work activities means that people are writing down what should be tracked automatically. Also, a proliferation of paper files recording work activities may indicate a CYA (cover yourself) mentality in your organization.

One side-effect of these problems is poor morale. People do not like making mistakes or performing mind-numbing, repetitive tasks. A process with any of the above problems puts workers in a system where mistakes are due to the system itself—beyond the workers' control.

Output Problems

In this context, output is not the product of a process (e.g., a delivered service or item) but refers to data generated by the process. Although output problems are not as deleterious to the quality of a process as bad input, the problems listed below are sure signs that the process is no longer meeting its original business objectives.

A large number of "management" reports: Although reports are often critical in decision making, a proliferation of reports is a sign either that the process is overmanaged or that the managers are not sure what to look for. On one project we worked on, a large dumpster was placed near the copy machine for recycling paper. Each day, the dumpster would be filled with unread management reports. How could anyone read a 1500-page report? Yet these reports were faithfully generated by the systems production support (and sorted, and distributed). All of this was a colossal waste.

Photocopying and storing of paper records: Despite the ubiquitous nature of computer support, most business environments still rely heavily on paper. In the design of several sophisticated systems we were involved with, one of the first operations requirements was to have a "Print Screen" capability. Why? So the user "could have a record of it." Such "records," in bulk, are useless. They typically are poorly cataloged and so are difficult to retrieve on demand. These paper records often have the same root causes that the paper logs described above have.

The best way to deal with these output problems is to take a green field approach. Start with the assumption that your new system will not generate any paper records or reports at all. Of course, people who are used to printing screens and storing them in folders will be hard to convince that this is a bad practice. But, for each request you receive, ask the requester what he or she will do with that information. How will providing them with paper yield a better product, or a happier customer, or reduce the cost or the time of delivery? Remind that person that there is a real cost associated with generating this information, and refuse the request if the requester cannot relate his or her requests to real benefits for the business.

Exchange Problems

Input and output problems are sometimes localized to individual systems. Perhaps the ordering system is old and particularly fragile, and so the users' requests for new features have not been fulfilled. However, most processes involve multiple organizations, each executing tasks according to its function and each supported by its own information system. Problems with exchanging data between work groups and between systems are more serious and are symptoms of a fragmented information architecture. This fragmentation creates islands of information, where each organization is isolated from the others because the systems support individual functions rather than the entire process. The problems listed below are evidence of the lack of a global system architecture as well as a poor understanding of how the various functions must work together in the extended business process.

People serve as organic I/O devices: We have seen skilled, intelligent workers spending most of their time taking data from one system and typing it into another, perhaps the single most inefficient and error-prone way of having computer systems exchange data. The problem is that the systems involved do not interface directly. Perhaps the systems were built at different times or by different organizations. However, it is often easier for managers to have someone on the payroll re-keying data than to pay for and plan the development of a machine-to-machine interface.

Most communications are via mail, fax, e-mail, and phone calls: If several people are involved in the execution of a process and they are not co-located, they try to share information by any communications media available to them. Thus, the result of having different functional organizations support a process is a flurry of e-mail, voice mail, and fax communications. None of these is timely enough or as beneficial as direct communication, and they are a cause of delay and error.

"Folders" tie work groups together: If no one system can support the process (or even serve as a gateway to the other systems), then workers will find creative ways to store the information they need. One way of providing "centralized" storage of information is to have every person involved in a process leave a paper trail and hand it to the next person to work with. In a way, this mechanism

can tie together disjointed work groups that may each have their own systems. However, it also forces the process to be strictly serial (only one person can have the folder at a time) and allows only very few people access to the information in the folder.

These problems typically are resolved by consolidating work groups and systems and by building machine-to-machine interfaces between systems. Of course, the best solution is to obviate completely the need for the exchange in your redesigned process.

Usage Problems

As you walk through the existing process, the last set of problems to look for centers around the interaction between the systems and the user community. The types of problems listed below arise when systems do not meet their original design or when the systems designers (including those putting requirements on the system) do not understand the real process needs.

Only a subset of the process is supported by systems: An extended process typically will have several functions that require information system support. If only a fraction of the people involved in executing the process are supported by a system, the people designing support systems do not understand the business process. System support for individual functional tasks within a process leads to fragmentation and consequent error and delay.

There is a proliferation of underutilized workstations: More than once, we have seen work areas brimming with hundreds of workstations (usually PCs) and observed that all of that computing power was rarely used. Clearly, the original designers anticipated more usage than this. Determination of the root cause of this situation (e.g., the systems do not work as planned, and manual methods are used instead) is necessary to avoid continuing the costly mistakes of buying unnecessary workstations.

There are many high touch-time tasks: High touch-time tasks are those that require the most manual effort. Workers who perform these tasks may often be bottlenecks in a process, and so can cause costly delays. Examine these tasks closely to find out why they are not better supported and how the amount of touch-time could be reduced or even eliminated in a new design.

As you walk through the process, you will encounter many input, output, exchange, and usage problems. The root causes of these problems will be an important input to your redesign.

THREE SCENARIOS ILLUSTRATING INFORMATION ARCHITECTURE PROBLEMS

This section is meant to provide another way of identifying the root causes of problems with the information architecture. While walking through the process, you have been building a paper model of the architecture, showing which systems are used and whether they exchange data. In examining businesses ranging from retail sales to telecommunications, we have found three common scenarios, each with its own root causes. In your own effort to understand the current information architecture, you will likely find what we coin a Tacoma Narrows Bridge, an Autoland, or a Staten Island scenario, named after other disasters that had the same root causes as these scenarios. If you recognize one or more of these situations in your own information architecture, you will have another means of identifying root causes so you can address them in your redesign.

The Tacoma Narrows Bridge in Washington (known to motorists who drove it as "Galloping Gertie") is familiar to all civil engineers as an example of how not to build a bridge. In 1940, four months after the 2800-foot span was built, the bridge collapsed because the engineers did not consider the effects of harmonics on the structural integrity of the bridge. *The Tacoma Narrows Bridge scenario applies when all of the needs of a business are not considered in the design.* For example, an information architecture that is not completely thought out also will fall apart. A typical case is illustrated in Figure 8-1, in which an on-line view of corporate resources is made available only to a subset of the business. The result is that the information is impossible to maintain and does not provide the business with the anticipated strategic advantage.

By far, the most common cause of the Tacoma Narrows Bridge scenario is that the systems are developed as independent entities, each serving a particular business function. The result is that there is a system for each organization. The people in each organization (i.e., each business function) then become experts in a particular computer system. Such a Humpty Dumpty architecture rarely has the pieces talking to each other in a standard way. Thus, to execute

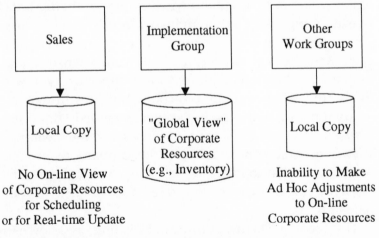

Figure 8-1. A Tacoma Narrows Bridge scenario.

any process that involves multiple functions, the operations staff must do what they can to keep the pieces together. This leads to the repeated data entry and the flurry of electronic mail, reports, and telephone messages cited above.

The Autoland scenario is a bit more subtle. The film *Roger and Me* popularized the story of Autoland, an amusement park/mall developed in Flint, Michigan. At the time, Flint was the home of a major General Motors plant, and the park was built with an automotive theme. Autoland was huge, with attractions that included an indoor roller coaster. Despite the relative obscurity of Flint, Autoland was expected to bring in three million tourists a year. It did not—in fact, it did not even come close. And so, after millions were spent on its development, Autoland failed to fulfill its grand vision, all because of a poor understanding of what the town of Flint needed coupled with the inability to create a major entertainment attraction.

The Autoland scenario describes individual systems built with a grand vision that was unrealistic for the business, given its resources and capabilities. We have seen a number of unfulfilled visions, each of which smacked of that Mickey Rooney brand of naiveté—"We'll put on a show in the barn!" For example:

- "We'll build a system to handle order fulfillment, implementation, testing, and billing."
- "We'll build a single shared database for all of the major functions within our business."

These visions presented major technical challenges of system integration. Each required supporting many users, a large database, and a wide range of user communities and system functions. In these cases, a poor understanding of the process (and a poor understanding of technology) led management to underestimate the task at hand.

When the complexities became apparent late in the development cycle, the original design was inadequate, and the organization did not have the technical skills required to create an adequate design. Each case was an Autoland scenario. Millions of dollars were spent, and the unveiling of these systems to operations users was met with little more than a yawn. Once again, management would have to find a way to get work done with inadequate tools until the systems organization tried again with the next vision.

The third scenario is the Staten Island scenario—Staten Island being the little-known fifth borough of New York City, which is famous for a ferry that takes people away from Staten Island (with a good view of the Statue of Liberty) and for a landfill. (This landfill is the only man-made structure that is visible from space besides the Great Wall of China, and it has more volume than the Great Wall.)

A Staten Island scenario describes a system that is poorly designed and is poorly integrated with the rest of the information architecture. After years of paying higher taxes for the luxury of storing New York's garbage and providing byways for New York's traffic, Staten Islanders are asking themselves why they are even part of New York City, and they are attempting to secede from it. With Staten Island systems, users and developers often ask their own versions of important questions *after* the system has been built. For example, "How do we synchronize what we store in our [independent] database with all of the other copies of the same data?" Figure 8-2 shows a classic case, where the resulting system stands alone, with no interfaces to other systems, very few users, and lots of data entry and reports.

Reading about these disasters is depressing, and we must ask, how do these things happen? How do large, otherwise sensible corporations produce such mishmashes for information architectures?

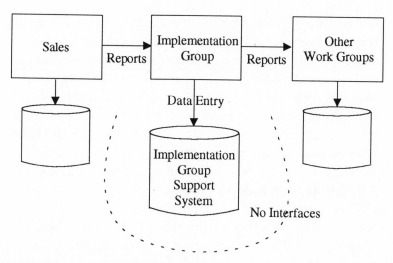

Figure 8-2. A Staten Island scenario.

A major part of the problem is the willingness of individual organizations to create their own information system support. Typically, an organization has a need that a manager thinks can be served by a system. Even if the manager knows something about technology, which is doubtful, he or she almost certainly does not have a global view of the organization and all of its supporting systems. All the manager knows is that there is a need, and that the MIS department said it would take a long time and a lot of money to support that need; and, because the MIS department is supporting the other organizations, the manager's need will have to be modified somewhat to fit in with the existing system.

With all of these roadblocks, the manager decides to go it alone and commissions some bright people in the organization who know something about development to create yet another system. This system is totally under the control of the individual manager—but how does it get its data? What type of database manager does it use? What hardware and operating system are used? The truth is that such a system is built by using whatever tools are known to the people in the organization, and generally the system is a complete "hack." It is just something thrown together that may meet the immediate need but is difficult to support and grow. Even in those cases where the system is well built, it likely does not fit in with the other systems supporting the business process. For example, if the MIS organization works solely on mainframes, it is doubtful that an independent development organization will have any idea of how to make its application work in concert with the other corporate applications. Thus, you wind up with the exchange problems and input problems mentioned above. From a global view, you wind up with another Tacoma Narrows Bridge.

EXAMINING EACH SYSTEM

As we have said, to understand the information architecture you will follow the process flow and also look at each system in detail. The goal of examining individual systems is to quantify their costs and to gauge their value to the corporation. Here, value includes the yearly cost of maintaining and growing a system, its inherent architectural quality, and the benefits derived from its use in support of the process.

Cost

In a typical business, few people understand the real cost of operating a computer system. People tend to relate system cost to the cost of the initial development and the purchase of the hardware. However, the ongoing yearly costs associated with the systems usually are more important than the initial costs. You will use this information to help quantify the benefits of your redesign. (This type of analysis should already have been done in the "discover" phase.) For example, if your new design calls for the consolidation of several systems, you will include the yearly cost savings of unplugging those systems to offset the development of the new system.

For each of the systems involved in your process, collect information on the amount of money spent on simply keeping the machine alive: hardware maintenance, renewal of software licenses, software upgrades, salaries for staff to support the system, and even storage, power, and cooling costs. Also, you need to calculate the cost of new development on the specific system. This is difficult because the cost depends on the particular application or feature that is being built. However, you should benchmark the costs of recent development efforts: a simple change (e.g., adding new data elements to a screen) and a full-fledged new feature. Note the time that each development takes. Long development cycles not only inflate the staff costs but also impose an opportunity cost on the business.

Robustness

The word "robust," when applied to a computer system, implies that the system is built in such a way as to make it easy to add capabilities or usage without completely redesigning the system or breaking unrelated functions that used to work. Although judging the robustness of a system is more subjective than using cost figures, it is important to know whether a given system is technically sound, and whether it will support future needs.

There are few hard and fast guidelines in this area. Certainly, one should answer at least the following questions for each system:

- Can I add or change features without breaking other features?
- Can I accommodate additional users and functions without having to rearchitect the system?
- Does the system exchange data with other systems using nonproprietary interfaces?
- Is our application-specific code the only custom software, or were standard functions (e.g., window managers, error loggers, code generators, database management systems) also custom-developed?

The robustness of a system is a consideration when you are deciding whether systems must be consolidated to support the new process.

Functional Value

Even if a system is well designed and inexpensive to operate, it still may not be of value to the corporation. The tracking systems example at the beginning of the chapter is a case in point. Even if each of those tracking systems were a paragon of systems development, the obvious overlap in function would have to be corrected. Despite the importance of tracking to the corporation, the functional value of an individual tracking system is limited to the single organization it supports.

Having examined how each system is used to support the process, you need to step back and ask, for each system, "What is the purpose of this system?" Does this purpose, when considered in the context of what the corporation is trying to achieve with the business process, make any sense? Does the way the system is used or the way it exchanges data with other systems support the overall process?

The answers to these questions, along with information on cost and robustness, will be used in the redesign phase when you are considering consolidating or eliminating systems.

Innovate and Build

Experience alone will not guide you through a reengineering project. In addition to your experience, you need a way to explore the consequences of possible new designs *before* you actually implement them, and you need tools to help you incorporate the positive elements of a reengineered business into your own design. This phase is devoted to these topics, and Chapter 9 introduces this phase with a discussion of preparing for the redesign.

Complicated things such as redesigning a business can seem uncomplicated when described by a familiar analogy. For example, John Zachmann describes this system development methods using a house-building metaphor.[1] First, the customer gives a broad outline of the house, then the architect develops the plans for the house, and finally the contractor implements those plans. This metaphor works reasonably well because house-building is an activity to which most of us can relate. People have been building houses for thousands of years, and most of the factors that are important in designing and building a house—materials, insulation, and so forth—have been thoroughly studied and documented.

The house analogy, however, does not serve reengineering projects at all well. Experience may tell the general contractor how ambient humidity and concrete drying time are related, but the whole point of reengineering is to put things together in *new* ways to achieve a competitive advantage. In spite of reengineering's popularity, there are few successful reengineering efforts. In creating innovative designs using advanced technologies, pioneering companies are pushing the envelope of corporate experience.

---

[1]Zachman, J. A., A framework for information systems architecture, *IBM Systems Journal* 28 (3):275–292, 1987.

METHODS AND PRINCIPLES

The output of Phase III is the redesign: the new process, a description of system functions, and an organizational structure, including who will perform which tasks in the new process. (The organizational structure will be examined in much more detail in Phase IV.) The worst way to develop your new design is to lock up your reengineering team in a room, brainstorm the new business, and then foist it on the rest of the company. Reengineering need not, and must not, be a paper exercise.

In our experience, there is a propensity for project planners to engage in hyperplanning. The house-building paradigm misleads people into believing that if they could only come up with the perfect blueprint, then the implementation would follow without a hitch. Wrong! You cannot judge the consequences of a given design by eyeballing it. As the design represents a new way of doing business, you will not have any data points (any experience) to guide you in understanding these dynamics. Even if you think that your years in business give you a "gut feeling" for how things will work, the people–process–system interaction is actually far too complex for the team to understand without some of the tools introduced in this phase.

Applying the house-building metaphor to the development of complex, dynamic systems has led to disastrous results. Lowell Jay Arthur, in his latest book on software development,[2] summarizes the problems that arise from this metaphor:

> The construction paradigm insists that we have *complete knowledge* of requirements and technology, that we have fixed stages with specific, defect-free

---

[2]Arthur, Lowell Jay, *Rapid Evolutionary Development*, p. 4.

products, and no unresolved issues. Experience has shown that this paradigm is the road to ruin—excessive cost and schedule overruns, employee burnout, and outright customer mutiny. System construction would work well if the requirements were both well known and static, neither of which is true of information systems.

This scathing review of the traditional systems development methodology applies to the development of *any complex system,* including work groups and business processes as well as information systems. Innovation—building something new—is inherently risky and error-prone. The best you can do is to know why you are including a particular element in your design and to spend your resources wisely in implementing that particular element. Chapter 9 describes things you must do to prepare for the new design. Chapters 10 and 11 will help you in a number of design areas:

- Dividing work among people and machines.
- Deciding when to mechanize work and how to marry newer technologies to operations flows.
- Determining which factors really affect the performance of a process.
- Determining the best among many alternatives.
- Validating aspects of the design before the design is complete.

Instead of first building a complete, complex specification of the new business, Chapters 10 and 11 describe a method for evolutionary development, of both the system and the new process. Chapter 10 contrasts an evolutionary reengineering methodology with a more traditional methodology and explores the consequences of using one or the other. Chapter 11 describes a structured means for deciding which elements should be included in your design and comparing alternative designs.

The remaining chapters in Phase III relate to the redesign itself. In the previous phase, we emphasized the need to identify the root causes of the problems in the current environment. Chapters 12 and 13 describe tools that can be used to address common root causes, such as data integrity problems and fragmentation of information system functions. Chapter 14 presents a list of principles, compiled over a number of projects, that we continue to apply to our redesign efforts. This list should serve as a reference (or, more to the point, a sanity check) throughout your redesign. Finally, Chapter 15 ties the methods and tools together in a step-by-step example of a redesign effort.

9

Preparing for the Redesign

An underlying theme of the "innovate and build" phase is that it is difficult to understand a priori how a business system—processes, systems, and organizations—will behave under different conditions. Thus, we advocate an evolutionary approach toward implementing the new design. In the last phase, you compiled root causes of the current problems, and you commissioned benchmarking studies and customer satisfaction surveys. The results of these exercises serve as the seeds of the new design. In this phase we describe techniques for building software and for making design decisions that will allow these early ideas to germinate. Before you actually begin exploring design alternatives, you must do three things to prepare for the evolution:

1. Clearly state the objectives of the new design.

2. Form an operations team to help explore design alternatives.

3. Select a portion of the production work for the team.

This chapter provides some heuristics for carrying out these tasks, and explains how each aspect of the new business system can be implemented in an evolutionary fashion. A more detailed handling of systems development and organizational change is provided in later chapters.

SPECIFYING THE PROJECT'S OBJECTIVES

A characteristic of successful people and successful companies is their ability to state clearly what they want to be and do. One effective way of doing this is to draft a mission statement. In *Seven Habits of Highly Successful People*, Stephen Covey recommends writing a personal mission statement to help you focus on what is important to you. The act of writing such a statement forces you to think about the principles that guide your everyday life.

Before you consider new designs, you need to think about the mission of your business and the goals of the process you are reengineering. Just like a personal mission statement, your project's objectives help everyone focus on what is important. If your goals are measured by the number of keystrokes someone can type per minute or the number of phone calls that a customer representative can handle per hour, then the people in the company will focus on those things. To avoid corporate myopia, the statement of objectives of a reengineering project should have the following characteristics:

- *It must be operationally defined:* Do not put words such as "quality" and "superior" in a statement of objectives unless you can translate those adjectives to the actual operation of your business. Quantifying how you compare to another company or to a previous benchmark makes the objective more palpable.

- *It must specify how and why as well as what:* There is no room for a slogan in a statement of objectives. In one company we worked with, banners proclaimed "Zero Defects Is Our Goal" even though defect rates were as high as 80 percent. "Zero defects" is an empty phrase unless it is accompanied by a statement of how to achieve it.
- *It must relate any measurements to the customer:* Your company is not in the business of improving internal measurements (keystrokes, number of phone calls); so these items should not appear in your statement of objectives. Instead, relate your objectives to the time, cost, and ease of customers' doing business with your company. Customers tend to care about how long it takes to get a product or a service, how much it costs, and how much pain they have to endure to get it. The employees supporting the process should have the same concerns.
- *It must provide a vision of the future:* Relating the objectives of the reengineering project to the vision of the company's future—its purpose—gives people on the project a sense of context. They can understand how their work is impacting both the short-term and the long-term success of the company.

Of course, many businesses have mission statements, but most of them do not have the above characteristics. They are, instead, Pablum for the public—lofty adjectives and truisms strung together, all of which say nothing about why a customer should do business with that particular company. For example:

> We believe in the free enterprise system, and we shall consistently treat our customers, employees, stockholders, suppliers, and community with honesty, dignity, fairness, and respect. We will conduct our business with the highest ethical standards.[1]

Mission statements and statement of objectives must not be so ambiguous. Customers and employees should be able to read the statement and understand what business the company is in and how it is going to be an even better company in the future. A good candidate for a reengineering mission statement is given below:

> Our purpose is to make organizations more productive by reducing delays in exchanging materials between locations. We are a package delivery service that aims to deliver packages of all kinds to domestic locations in less than 12 hours and to any point in the world in less than 24 hours. Within the next two years, we aim to reduce our average domestic delivery time from the current 18 hours to less than 12 hours for 95% or more of our packages. To do so, we will employ technologies to automate simple tasks, and our personnel will work closely with our customers to avoid known delays in pickup and delivery.

Such a statement tells the reader what to expect from the reengineering project as well as where the project fits in with the long-term business strategy. Once the reengi-

---

[1]This is one of the "corporate creeds" cited in *Corporate Philosophies and Mission Statements* by Thomas A. Falsey.

neering team and upper management agree on such a statement of objectives and corporate mission, it is time to assemble the personnel who will take part in the design and implementation of the new business.

Given the complexity and the cost of reengineering projects, most companies elect to run a pilot or a trial in order to minimize risk. If the new design does not provide the expected benefits, or if there are unforeseen problems, the project can be aborted before the entire company is reengineered. This strategy fits in well with the evolutionary approach we describe in this phase. To begin exploring the design possibilities, you will need to form a team to help realize those designs in a production environment, and you will have to give them work to do in order to test the team members' ideas.

FORMING A PROTOTYPE CASE TEAM

One of the principles of reengineering[2] is to organize people around extended processes instead of tasks or functions. When you have selected the scope of your project, you have identified the extended process. Now, you need to organize into one team the people who support the functions within that process.

In the 1980s, the Index Group wrote about multifunctional teams formed to support extended business processes. These "case teams" were so named because the entire team is responsible for the output of an extended process, not just a function within a process (i.e., they work on a particular case from beginning to end). Thus, an order

---

[2]General principles for reengineering projects are provided in Chapter 14. In *Reengineering the Corporation,* Michael Hammer and James Champy describe a number of organizational principles they formed, based on their experience with reengineering projects.

fulfillment case team might be responsible for creating the order, reserving products from inventory, and dealing with back orders and cancellations. Whenever possible, there is no hand-off of the order to any other work group. The team is responsible for the order from creation through its complete processing. If there are problems with filling the order, a specialist may be called in to help the team, but the team retains its responsibility.

As mentioned earlier, one of the tenets of concurrent engineering is the formation of a multifunctional team. Such teams have been successful in a wide range of processes, including developing new products,[3] delivering service,[4] and manipulating information.[5] Forming one or two such teams—prototype case teams—is the first step in the development of a new business design.

When you perform a trial of a new design prior to full implementation, you will necessarily have to manage a hybrid operations environment. Some work groups will continue to execute the old processes while the case teams are evolving the new design. You can do several things to ease the conflicts between the old and the new environments:

- *Ask for case team volunteers:* If you hand-pick the best of the operations staff, you immediately stigmatize the team, and you may have resistance from the people you selected. By asking for volunteers, you are assured of getting team members who want to be part of the new design.
- *Place all of the case team members in close physical proximity:* A list of names on an organization chart

---

[3]For example, Kodak successfully reengineered its product development process for disposable cameras.

[4]For example, Bell Atlantic successfully reengineered its installation process for high-capacity telecommunications facilities.

[5]For example, Mutual Benefit Life successfully reengineered its policy application process for life insurance.

does not constitute a team. Teams are formed when people work together toward a common end. "Together" means sharing coffee breaks, leaning over the desk to answer a question, exchanging personal likes and dislikes. Team members need not be close personal friends, but close contact is required in order for the project to benefit from small-team dynamics.

- *Separate the case team from the existing work groups:* New ideas often are unpopular. If the case team is not separated from the existing work environment, there will be subtle or perhaps blatant pressure to conform to the old way of doing business. Those creating the new design will necessarily threaten those who continue to do things the old way, and such an environment is not conducive to innovation.

- *Communicate the project's objectives and status to everyone:* Rumor and gossip will fill any void due to a lack of information. If you do not communicate openly about the objectives and the status of the reengineering project, you will be inviting the cynicism of people who are not directly involved ("I've heard they're going to downsize again"; "They're not doing anything differently—I don't understand what the big deal is").

As discussed in the "discover" phase, the leader and the public relations person (whose roles are defined in Chapter 3) must actively communicate project objectives and status. Being open with all employees who may be impacted by the project will make it easier for the case team members to interact with their peers in the existing groups. If the team members are well informed, their interactions with their peers can help mitigate negative reactions from those not directly involved with the project.

SELECTING WORK FOR THE TEAM

In a hybrid operations environment, work will be handled differently according to who is handling it, whether it is the prototype case team or the existing work groups. As most businesses are organized in one of three ways—by geographic region, by product or service, or by customer—there are three ways to segment the work for the prototype case team. In our experience, you can minimize problems due to a hybrid operations environment by segmenting the work by customer.

Managing a hybrid environment is inherently difficult and confusing, but segmenting by customer is the best approach. For one thing, it allows you to present a single interface to the customer. If you segment by product or by region, then customers who order different products or who order across different regions will be dealing with different work processes.

STARTING THE DESIGN PROCESS

With an evolutionary approach, the people working on the new design must be able to address all aspects of the business system—processes, systems, and organizations. Thus, there will be a need for dedicated technology personnel in addition to the reengineering team and the prototype case team(s). The technology personnel will work closely with the other teams *throughout* the redesign process in order to design the new information architecture and to develop any needed software.

The redesign process, then, is an interwoven evolution of the three facets of the business system. Initially, the case team will be working with the extended process you identified in the "discover" phase, including the existing information system support. *This does not mean that you will not make radical departures from the current environment;* rather, this strategy is meant to minimize the chances for failure at this critical early stage of the project. Even if you begin with the existing process and systems, you will encounter problems in several areas:

- Resolving personal and union issues in setting up the new team (and new work assignments).
- Resolving technical and administrative problems in co-locating the team members and in moving their equipment (e.g., terminals, communications lines).
- Providing information and direction to the new members of the project, such as the technology personnel.
- Communicating the mission of the reengineering project to upper management and then to the rest of the affected organizations.

Trying to begin with new systems, new processes, *and* new work teams will only lead to frustration and disappointment.

You may wonder how you can achieve radical improvements if you start with elements of your current environment. The key is working in the "laboratory environment" with the small case teams and a flexible work environment. Such an environment is more conducive to major systems enhancements, for example, than are traditional environments. Given the appropriate communications to the rest of the organization, they will understand the experimental nature of the new environment and will understand that not everything will work perfectly. With realistic expectations and with no rigid organizational boundaries or administrative policies in the way, changes can be made easily.

Both to demonstrate their ability to take action and to illustrate their support, the project personnel should begin immediately with small, short-term improvements. Early successes will help build confidence in the project and will facilitate teamwork. Means for introducing more significant changes are described in further discussions in this phase. The evolutionary introduction of technology is described in detail in Chapter 10. Process changes will be based on the data you compiled in the "hunt and gather" phase, including the root causes of current problems. Means of deciding between alternatives are discussed in Chapter 11. An extended example is provided in Chapter 15.

Although a laboratory environment does provide flexibility in introducing process changes and system enhancements (or even completely new technologies), it still is not appropriate to make organizational changes on an ad hoc basis. Mechanization of a mundane task can be given a trial for a few days and easily abandoned if necessary. However, such flux is not possible in dealing with personnel. Thus, we recommend forming the proposed team at the outset and leaving that team essentially intact for the duration of the redesign process. Simulation models may suggest more optimal personnel combinations when more information is available, but changes to the team structure should be made slowly and with the involvement of all of the team members. This topic is discussed in more detail in Chapter 17.

10

The Need for Evolutionary Software Development

Complex systems take a long time to build. Whether the system is an oak tree, a human brain, or an airline reservation system, it is not possible just to think about its design and then draw up a complete specification. That is not how nature works, and that is not how systems development should work. Ever since Darwin's finches, we have known that it usually takes a long time and many trials for a successful complex system to be created.

Like any process, the software development process often suffers from problems identified in the "hunt and gather" phase. Fragmentation, functional organizations, long delays, and rework all plague typical software houses. In this chapter, we describe traditional software development methods and explain why they are inadequate. We then present an emerging approach—evolutionary development—and show how it can be used during the "build and innovate" phase of reengineering projects. The techniques described in this chapter represent a breakthrough, in that process problems and system problems now can be solved concurrently. This is the single best way to apply technology to implement a new business design.

THE TRADITIONAL APPROACH

Most of the development organizations we have seen have used an information flow such as the one shown in Figure 10-1. Customers talk to systems engineers or

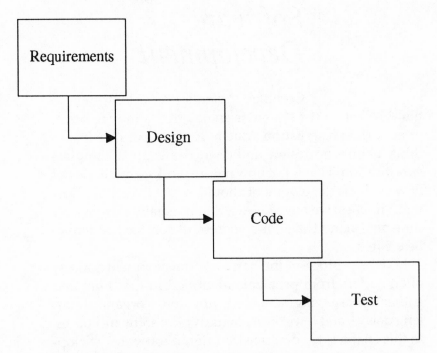

Figure 10-1. Traditional systems development information flow.

"analysts." (Throughout this chapter, the word "custom-ers" is meant to include both users of the system and the people or organizations funding the development.) These are people who may not know much about a customer's business or about information systems but have some technical orientation, and they serve to translate the cus-tomer's fuzzy statements of want into a neat requirements document. The systems engineer gives the document to a body of developers who try (from scratch) to understand both the original problem and the solution encoded in the requirements. When the developers have built something, their code and the requirements are handed to testers, who also do not understand the original problem, but who write another document that includes, in essence, a restatement of the system engineer's requirements. The tested software is given to a delivery organization, which consists of people who are versed in integrating the soft-ware and the hardware into the existing production environment. Finally, after a lengthy period, the customer gets it and sees just how well the original intent was translated (several times) into a working system.

This process, often referred to as the "waterfall method," is fundamentally flawed. Like any other process where many people are involved and those people are function-ally aligned, this type of software development process is slow, error-prone, and inflexible.

In the book *Prototyping,* Budde et al. summarize the problems with the traditional development life cycle with six theses, each of which is paraphrased below:

1.   *A complete and permanently correct description of infor-mation systems is not possible.* Few customers understand what computers are capable of, or how their needs can best be served by computers; instead they belong to the "I'll know it when I see it" school. Nevertheless, we spend millions on tools intended to meticulously note every aspect of a system before a customer has even a hint of how it actually is going to work. Further, the system must develop over time, *even during its development,* as users' needs change. The traditional software development process is aimed at "freezing" requirements and code as early as possible and then making changes as difficult as possible. (For example, there are typically change control systems, change review boards, change measurements, etc.) Freezing the specification does not change the dynamic nature of the business environment.

2. *Specification and implementation of a software system cannot be performed as separate work steps in time.* Although the goal of traditional requirements is to be technology-independent—that is, completely abstracted from the implementation—it is the implementation that helps identify inconsistent or unrealistic requirements. Clearly, the specification is necessary for the implementation, but the implementation helps make the specification more precise and truly technology-independent.

3. *Formal, nonexecutable specifications are largely unintelligible to users and developers.* In our experience, customers rarely read (and almost never fully understand) lengthy requirements documents, either textual or graphic (e.g., data flow diagrams). Despite very officious inspection and sign-off procedures declaring acceptance of the requirements, few people if any understand what the real implications are until the system is delivered to the field.

4. *Traditional development strategies exclude users from crucial activities.* This is perhaps the most insidious problem with the traditional approach. System engineers, at the beginning of the project, may talk with users (or, more typically, "user representatives"). Once they believe they have captured the requirements in a document, and after the perfunctory review and sign-off, they hand off the document to developers who proceed with the implementation. The user is almost never consulted on critical design aspects, such as the utility of the user interface design (menus, forms, etc.) in the operations environment. Some progressive developers may demonstrate a number of design aspects to the users, but this demonstration software is never truly exercised doing real work. These demonstrations do more for public relations than for the usability of the design.

5. *The label "maintenance" conceals vital aspects of development work concerned with adapting a system to the application context.* Maintenance usually is intended as a phase where errors, or bugs, are fixed. However, bugs usually stem from poor requirements. Thus, functions that should have been part of the original development are relegated to maintenance releases, where quality assurance and careful evaluation may be at a minimum.

6. *Life cycle plans are unsuitable for controlling software projects; milestone documents are produced for the benefit of management only.* This is, by far, our favorite of the six theses, for it succinctly captures the inadequacy of project management techniques for most software projects. Too often, the managers of software development organizations know too little about the business problem—and about information science—to manage software development successfully. Instead, they try to translate the project into that common denominator of all projects—a Gantt chart. It does not even matter if we are solving a business problem for a cost that is less than the benefits. Have we met our milestones? Did we review the requirements document last week as planned? These are the concerns of the traditional software development manager. In a review of software projects developed by using the traditional approach,[1] most projects had a product-oriented goal, not a task- or result-oriented goal.

As early as 1979, a widely cited study[2] concluded that a root cause of the failure of software projects was the inability of users to describe their requirements accurately and completely. The response to this study ranged from formalizing the specification part of the development process to providing specifications via an experimental, sometimes exploratory, use of working software prototypes.

Structured analysis and the use of Computer-Aided Software Engineering (CASE) tools represent one means of formalizing users' requirements of the system (these tools are described in some detail in Chapter 13). However, many CASE systems actually impede change in the development cycle. With structured analysis, a great deal of time is spent in documenting each detail in the specification. Because changes during the development process may mean reworking or completely redoing the structured analysis output, there is a strong tendency to keep things the way they are to avoid laborious documentation.

To many, structured analysis and CASE tools *embody* the traditional approach and do not support more dynamic approaches:[3]

> Despite growing sales figures for CASE systems and an aggressive marketing strategy, we feel that, for the foreseeable future, there is little that is likely to emerge from this quarter in the way of suitable approaches for

---

[1]Gilb, T., Evolutionary design versus the waterfall model, ACM SEN, 10:49-61, 1985.
[2]General Accounting Office (1979). *Contracting for Computer Software Development*, General Accounting Office Report FGMSD-80-4, September.
[3]This in not a universal position. For example, John Crinnion's *Evolutionary Systems Development* espouses the use of prototyping within a rigorous structured systems methodology.

supporting the evolutionary development strategy we advocate.[4]

What does all of this analysis lead to? The traditional approach is not suited to the current competitive operating environments. Customers need information system support developed quickly and cheaply, and it is difficult to respond to sudden opportunities with reams of documents and "change control."

In a reengineering project, the need to abandon the traditional approach is particularly pronounced. Where there is incomplete knowledge of the details of the redesign, it is necessary to test concepts while they are still in their infancy. Evolutionary development puts less emphasis on documentation and milestones than on delivering functions to the customer.

EVOLUTIONARY SOFTWARE DEVELOPMENT

Instead of aiming for a complete specification before proceeding, evolutionary development seeks to get *some piece* of the solution (e.g., support for a task or subprocess) working as early as possible. Lowell Jay Arthur provides the following definition:

> In rapid evolutionary development, prototypers create a basic working system which does not contain all of the infinite variety the customer ultimately desires, but which does work and provides the essential initial elements of the system. . . . Once this basic working

---

[4]Budde, R. et al., *Prototyping*, p. 146.

system is installed and turned over to the customer, a series of step-wise improvements—evolutions—turn the system into the customer's desired Garden of Eden.[5]

There are three main benefits to building systems this way:

1. You are more likely to deliver what the customer intended. System function and design concepts are demonstrated and validated early in the project.

2. The benefits of the systems are realized relatively early.

3. There is greater flexibility in dealing with new or evolving customer needs *during* the system development process than there are other approaches.

Remember the 80/20 rule of thumb? The rule states that 80 percent of the benefit typically comes from 20 percent of the product. A study published by Ernst & Young in 1989[6] concluded that "80/20 solutions . . . have a great deal to recommend them—80 percent of the ideal result achieved through *20 percent of the effort* that might have been expended. Companies can gain strategic advantage . . . through 80/20 solutions . . . ."

This 80/20 rule is implicit in common expressions such as "getting the biggest bang for the buck" or "picking the low-hanging fruit." Evolutionary development makes use of the 80/20 rule by focusing on delivering functions that may be simple to develop but will provide great value to

---

[5]Arthur, Lowell Jay, *Rapid, Evolutionary Development*, p. 56.
[6]Ernst & Young, The landmark MIT study: management in the 1990s.

the customer. Why should you withhold that value while the rest of the system is being developed?

Your reengineering project also should make use of the 80/20 rule. As you will see in the redesign example in Chapter 15, you will generate redesign candidates in an iterative fashion. If you can introduce these ideas into the field, then you can validate elements of your redesign as you go along. This approach reduces the risk of an all-or-nothing transition and starts to show some of the desperately needed benefits of the reengineering project.

Imagine that you are part of a team responsible for reengineering the joke delivery process of Comic Delivery, Inc. (whose current slogan is "We will serve no line before its time"). The CEO wants to change the corporate thrust by providing speedier service and billing, and so develops a new slogan ("Comic Delivery—a laugh a minute") and commissions your team to redesign the delivery process, including order entry, tracking, delivery, and billing. Now suppose that your redesign includes new technologies (hand-held computers, expert systems, etc.), a large, distributed database, and more than a half dozen interfaces to existing systems.

Using the traditional approach, you will most likely develop a complete specification of the redesign before you do any significant field-testing. Of course, you may conduct a trial of the design at a pilot location. You may have only twenty of 200 comedians using the new process, or you may only include one-liners and leave anecdotes out of the trial. Nevertheless, the traditional approach focuses on testing the *complete* design (i.e., all of the interfaces, all of the technologies).

Using an evolutionary approach, you will use the results of your analysis (see Chapter 15 on using a simulator to rank design alternatives) to identify the "low-hanging fruit" and to conduct a trial of that element of the design as soon as possible.

HOW TO DO IT

Programmers have been using an evolutionary approach for decades. For example, developers have long built prototypes to get customer feedback and to demonstrate system functions. What is new, however, is a change in technology that allows developers to build software quickly *on production-grade platforms.* Technology improvements of the last few years have made unnecessary the assumptions underlying the traditional approach. Now, instead of spending person-years writing detailed specifications and test plans and having change review meetings, the systems personnel can do what they do best—develop systems.

The evolutionary development method uses rapid prototyping to deliver working software to a production environment as quickly as possible. In the past, developers built disposable prototypes, which might include PC-based demonstrations with a small sample database, and which use a different architecture from the target solution. While this is useful for showing users what might be developed, it typically does not allow users to interact with the system as they normally would (i.e., talking to external customers and working orders). Hence, the word "prototype" usually connotes a toy development project that has no practical use.

It need not be this way. "In Genesis, God didn't create a prototype, show it to the users, and then tell them they would have to wait 15 million years for the real thing."[7] Similarly, you should not develop some tantalizing demonstration software only to tell the users that they can get the production version in six, nine, or twelve months.

---

[7]Arthur, Lowell Jay, *Rapid, Evolutionary Development,* p. 55.

Lowell Jay Arthur perhaps sums it up well: "Disposable prototypes have not worked." And there are two main reasons why not:

1. Disposable prototypes usually do not provide any real benefits to the customer.

2. When they do provide some benefit, sometimes they are not disposed of properly. If the user thinks the "real" system may not be delivered, or if funds for the project have dried up, then the operations personnel will cling to the prototype, and the information architecture is patched yet again.

The solution is to build the prototype using the target hardware/software architecture and to develop the application software using tools that permit rapid development. Two converging trends make this possible:

- Hardware is becoming cheaper and more powerful.
- Software is becoming more flexible and more powerful.

Software that once required the computing power of mainframes now is routinely executed on workstations and PCs. Moreover, new tools such as higher-level programming languages and interactive, integrated development environments have improved programmer productivity more than tenfold. This combination has changed the nature of software development and made it possible to evolve prototypes on a production platform.

Two technologies in particular, client/server architectures and application development tools, have made it much easier to practice evolutionary development. These technologies are discussed in detail in Chapter 12, but an example is provided here to demonstrate the feasibility of rapid prototyping.

Suppose Comic Delivery, Inc.'s reengineering team identified a root cause of delay to be that comic material is stored in several different databases on several different systems. A customer might ask for a Dan Quayle joke, and the operations staff would have to perform a (long) search of the one-liner database, the anecdote database, the pun database, and so on. Having users log on to each system and search for specific material is both time-consuming and error-prone. Now, the team's ultimate design includes a single, consolidated database, but this will take months of planning and negotiation before any development can begin. Instead, using an evolutionary approach, the team's developers decide to build a single front-end to the multiple systems, as this will alleviate the manual interaction with each system and provide a single, simple interface to the users.

Each of the existing systems is built on a different platform, including a different operating system, and this makes even a limited integration project difficult. However, the new system is based on a client-server architecture, which allows developers quickly to link in new functions as needed. The developers first build simple (but crude) interfaces to the existing systems using terminal emulation, and they use a screen generator to develop the user interface. For more than a decade, developers have built user interface demonstrations in a matter of hours using screen generators. The problem has been that these screens would be stand-alone, isolated from the rest of the application. Screen generators now have advanced so that the generated code includes instructions for communicating with other processes in the system.

The first attempt at a prototype is a single graphical user interface that performs commands on remote systems by emulating a user logged onto those systems. If, based on user interaction with the prototype, this is determined to be a good thing, then the next evolution might replace the interfaces one-by-one with more robust interfaces. With a client-server architecture (described in Chapter 12), a server that talks to the anecdote system using terminal emulation could be replaced by a server that uses a machine-to-machine interface (or even remote database accesses) without changing the other pieces of the system. The function of providing data from the anecdote system remained the same (i.e., the same data were retrieved), but the implementation of the server was changed. Such a change is completely localized in a client-server architecture.

An evolutionary development approach allows you to proceed step-by-step toward your ultimate design instead of carefully planning the trip and trying to make it in one great leap. This approach is essential to successful reengineering. Because an evolutionary approach is more flexible than traditional approaches, it can more easily support changes in the business environment and will lessen the current dependence on turnkey systems (i.e., single applications that are intended to meet the needs of many customers). Also, because an evolutionary approach marries the solving of business problems with the development of the solution, it makes a lengthy requirements analysis period unnecessary.

IF YOUR COMPANY ONLY CAN USE A TRADITIONAL APPROACH

Perhaps you cannot use an evolutionary development approach. Your systems organization itself may need reengineering. Then proceeding with the redesign will increase your risks as well as the time that it takes to realize any benefits of the project.

If you cannot incrementally test the elements of your redesign, you must rely heavily on simulation for your testing. The model of the existing process that was described in Chapter 7 will serve as the basis for a model of the new process. There are two distinct benefits to doing this:

1. You will be forced to think about and quantify the assumptions underlying your new design.

2. You can preview the dynamic behavior of your new design and make needed adjustments.

As you proceed with the redesign, you will answer the same questions you had to answer to build the original model. How long will this task take (at a minimum, at a maximum, and on average)? What will the arrival rate of orders be? How many people of which job functions will be executing the process?

Some people react with skepticism when we describe the benefits of building a model of the new design as a means of "testing" prior to implementation. We often are asked whether a model of the new design is at all valid if the design is not based on the current process but is built from scratch. The answer is clear when you examine the alternatives. If you use the traditional approach and do not build a model of the new design, then you are left only with your "intuition" and "experience" to guide you.

A model is not magic, but how else will you estimate the effect of automating a task or reducing staff? What will happen to throughput and unit cost? Will you eliminate bottlenecks or create new ones? If you cannot actually implement pieces of the design and measure the effects, then the next best thing you can do is model *what you assume will happen.* Then, use the power of the computer simulation to run thousands of orders (or tens of thousands) through your new design, and let the model unfold dynamics that you would never be able to conjure up on your own.

Clearly, the model of the new design will be more credible if your new design contains elements of the old process (for which a model was already created and validated). However, whether or not your design is completely new, simulating the new design is the next best thing to actually delivering pieces of the design to the field.

11

Traveling through the Process Space

Most of the managers we have dealt with employ an ad hoc approach to designing a business, relying solely on their experience and intuition to organize their people and design their work processes. These managers make major changes to their operations—consolidating work centers, introducing large-scale information systems—without any real understanding of whether the changes will make a difference in the company's performance.

Businesses, however, are complex systems, and people generally are ill equipped to understand the behavior of such systems. Whether it is an order fulfillment process or an ecosystem, people have difficulty understanding how elements of a complex system interact. Thus, they cannot know how to change part of the system to achieve a desired effect.

In this chapter, we introduce techniques to compare alternative business designs in a structured way. Combined with computer simulation, these techniques raise decision support to a new level. There are complex mathematics that support these techniques, but the concepts actually are quite simple, requiring an understanding of the business domain and a background in applying statistical methods.[1]

Our intent in this chapter is not to teach you everything about 2^{k-p} factorial designs[2] but to show you that there are proven methods you should use to aid your decision making as you create your new business design. Many of the techniques in this chapter have been used in manufacturing and chemical environments, but only recently have they been applied to business process design. The experiments we describe are not those of some sterile laboratory. Instead, they are meant to be conducted in the heat of business operations. The experimenters are meant to work closely with the personnel executing the process so they can become intimately acquainted with the work they are modeling.

WHY SYSTEMS ARE COMPLEX

In the early 1900s, a pestilence was destroying Australian crops. Scientists seeking to eradicate the pests had attended a conference in Hawaii where they learned that horny toads were a natural enemy of the bugs devouring the

---

[1]Dr. Jim Pennell, who introduced us to the material in this chapter, had more than fifteen years' experience in applying computer simulation and statistical methods. Other consultants in this area, such as the professors at the Center for Quality and Productivity Improvement at the University of Wisconsin, employ these techniques in a wide array of corporations.

[2]For more information on the methods described in this chapter, consult the related materials in the Suggested Reading section.

crops. Happy with such a simple solution to their problem, the scientists introduced over a half-dozen toads to the Australian pests. The initial results were terrific; to the delight of local farmers, the toads adapted well to their new environment and devoured bugs in impressive numbers.

Soon, however, the simple solution proved not to be so simple. The toads liked their new home all too well, and, lacking any natural predators in Australia, they multiplied in record numbers. The farmers' delight turned to dismay as the toads became a menace, covering roads and backyards, creeping into kitchens and bathrooms. The toads were everywhere, and they remain to this day, symbols of the difficulty people have in understanding the behavior of complex processes.

A more process-oriented example is provided by Peter Senge in *The Fifth Discipline*, in which he describes the "beer game" to demonstrate that people have difficulty understanding the dynamics of a process even when simple changes are made. In the beer game, people attending Senge's seminars play the roles of a grocery store manager, a beer wholesaler, and a brewery manager. Each person manages his or her own ordering and inventory, as imaginary customers buy beer from the grocery store. Figure 11-1 shows both the customer ordering activity and the inventory of each person over time.

Senge has played the beer game hundreds of times, yet the results are usually the same. The customer orders simply double at one point (during a promotion). Yet, because of the delay inherent in receiving the beer that is ordered, the grocery store manager continues to order more and more beer. The beer wholesaler, who only sees the orders of retail stores, thinks that there must be a tremendous surge in demand, and so he orders increasingly more beer from the brewery. Similarly, the brewery sees an upswing in orders from wholesalers and so increases its capacity. The game lasts longer than the delay in the

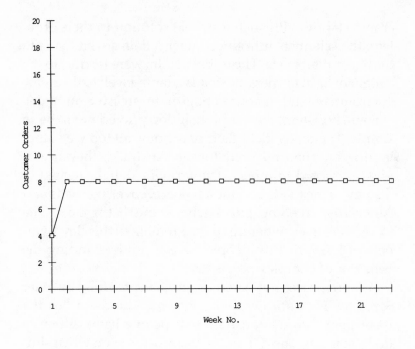

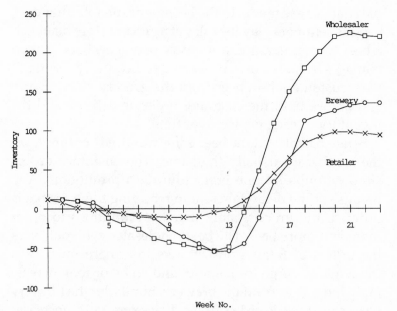

Figure 11-1. Results of the Beer Game.

process, so that the original orders (those at the point of increased customer demand) and those in the next few weeks are all filled. At the end of the game, each person invariably is left with a large inventory and little demand. (*The Fifth Discipline* catalogs dozens of such systems in which people have difficulty understanding the system dynamics.)

EXPERIMENTAL DESIGN

People have long struggled to understand system dynamics in order to change system elements to effect a desired result. In the 1600s, the scientific method paired structured observation with directed experimentation and revolutionized the way people understand systems.[3] Structured observation is carried out by using the tools and methods described in the "hunt and gather" phase. However, observing a system or a process is not enough; to ensure that the new design meets the objectives you set, you must use directed experimentation.

R. A. Fisher, a statistician, developed statistical experimental design in the early 1920s. In 1919, Fisher worked with farmers to find the best way to grow their crops. The traditional methods dictated controlling all variables except one in the crop-growing process and then measuring the effect of changing that one variable. Then, a different variable would be changed with all other variables held constant, and those effects would be measured. There were three problems with this approach:

- It was slow. Finding the effects of the different variables in the crop-growing process while holding

---

[3]The role of the scientific method and experimental design in quality improvement is described in an excellent article, "The Scientific Context of Quality Improvement" by George Box and Soren Bisgaard, which appeared in *Quality Progress,* June 1987.

other variables constant would take many years, as each round of experiments required full planting and harvest seasons.

- It was difficult to control all variables. Other than using an artificial environment, there was no way to control temperature, humidity, and so on, from experiment to experiment. And, if an artificial environment were used, there would be no way to predict whether the results applied to normal (i.e., variant) conditions.
- It did not measure the effect of interaction between variables. By definition, if you vary only one factor at a time, you cannot identify the effect of changing multiple factors at one time.

Using the traditional approach, discovering the effects of just three variables might take twenty-four experiments: four trials (for reliability) of experiments using two different values (a high and a low value) for each of the three variables. Instead, Fisher decided to vary all the factors simultaneously using a factorial design. With such a design, discovering the effects of three variables *and the effects of interactions between the variables* would take just eight experiments. Figure 11-2 shows a two-level, three factor designed experiment.

Each point in the cube in Figure 11-2 represents the results of an experiment with one of two values for each of the variables (2 values * 3 variables = 8 experiments). Despite performing only one-third of the total number of experiments, this design has the same amount of replication as the traditional approach; that is, there are still four experiments with the low value of Variable 1 (the bottom of the square), four experiments with the high value of Variable 1 (the bottom of the square), four experiments with the low value of Variable 2 (the left side of the

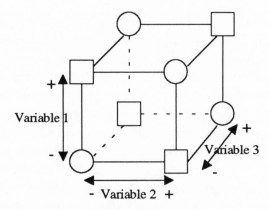

Figure 11-2. A factorial experiment.

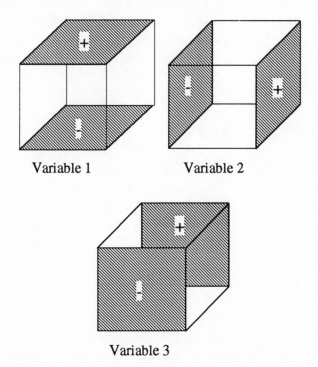

Variable 1

Variable 2

Variable 3

Figure 11-3. Traditional replication captured in a single experiment.

square), and so on. Thus, the results of having a variable set at a given value still are replicated four times. Figure 11-3 illustrates how this design still has four trials in which the change of one variable is isolated. Each cube in Figure 11-3 shows the effects of changing a single variable. This design also quantifies the interaction effects. By calculating experimental results across points in the cube, you can measure the effect of changing multiple variables (e.g., the interaction between Variable 1 and Variable 2).

These techniques at first may seem complex and even counterintuitive; but, despite the complex mathematics underlying these methods (see the Suggested Reading for more information), Fisher was able to carry out these experiments with farmers living near a small agricultural research station outside of London more than seventy years ago. There is no reason why, given the proper training, these techniques should not be useful to us in comparing design alternatives.

THE PROCESS SPACE

What has made directed experimentation a useful tool over the last seventy years is that elements of nature usually are related in predictable ways. The same is true for most business processes: as you vary an individual element, the changes tend to follow a continuous, smooth pattern. Consider the process of making your favorite dish (say, Stepper Chicken[4]). There are many elements or factors involved in the process: the amount of time in the oven, the temperature, whether or not eggs are used, the amount of other ingredients (marjoram, white wine, etc.). If you want to optimize a result of the process, often called

---

[4]The recipe is included in the appendix.

the "response," you may change one or more of the process factors. ("Next time, I'll try it without eggs.") Based on your understanding of the problem domain, you select the factors you believe to be significant as well as the high and the low values for each factor (some factors may have values of "Yes" and "No"). If you carry out these experiments in a structured way, you can make assertions about the effect of changing factors on the response. For example, Figure 11-4 depicts the effect of changes in temperature (all other things being equal) on taste. Given an oven time of 30 minutes, no eggs, 1 teaspoon of marjoram, and 1/2 cup of white wine, there is indeed an optimal temperature at which chicken should be cooked.

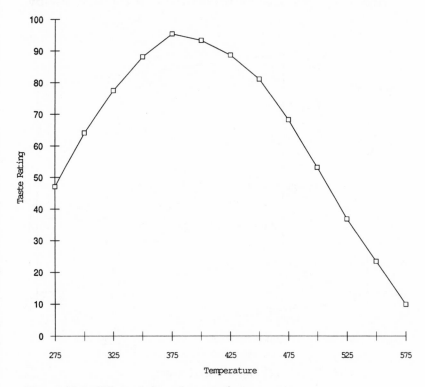

Figure 11-4. Effect of changing one factor.

Note that the relationship between temperature and taste is nonlinear; that is, a 25 degree change in temperature from 300 to 325°F has a different effect on taste from that of a change from 500 to 525°F. If the relationship were linear, the graph in Figure 11-4 would be a straight line.

We could also study changing two factors at the same time. The effect of changes in the two factors also could be plotted, this time with a three-dimensional rather than a two-dimensional graph (see Figure 11-5).

Notice that plotting the effect of all the different combinations of two factors produces a landscape of sorts. The optimal combination occurs when a positive response (taste, in our example) is maximized. Looking closely at Figure 11-5, you see several hills and valleys in the graph. These correspond to local maxima and minima; that is,

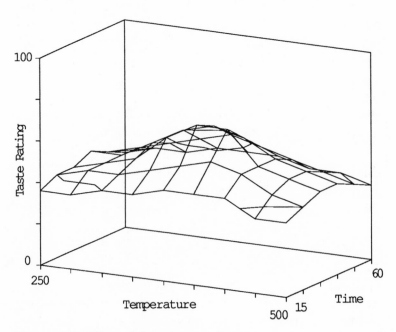

Figure 11-5. Effect of changing two factors.

those points in the graph represent combinations of the two factors that yield good and bad responses relative to other combinations in the neighborhood of those points.

As you consider the effect of changing more factors, you are adding dimensions to the landscape, producing the "process space." Whether you are cooking or delivering goods and services, the range of alternative process designs forms a space. It is possible to design experiments to discover the shape of this space and find the local maxima and minima—the best and the worst process designs. Without experiments, your understanding of the problem domain will give you only a rough sense of how the process space is shaped.

Fortunately, the concept of seeking out the hills and valleys in an N-dimensional space has been employed in many fields. In physics, it is used to determine the minimum energy required for a given set of particles. In designing neural networks, it is used to find out which networks will perform best for a given class of problems. In each of these fields, the N-dimensional space represents a range of alternatives. Again, experience and a knowledge of the problem domain are used to select the factors. Then, proven statistical methods are used to traverse the space to find the local optima.

Consider rolling a marble on a model of a three-dimensional landscape. You release the marble from some point on the landscape, and the laws of physics determine the path the marble will take. When you are considering the effects of multiple factors in designing a new business process, there are no natural laws to guide you, but there is a set of statistical tools that can direct you to a local optimum. The application of these tools to find an optimum is called the "response surface methodology." The response surface is the N-dimensional landscape formed by the different combinations of factors. George Box and

Norman Draper describe this methodology in their book
Empirical Model-Building and Response Surfaces:

> Response surface methodology comprises a group of
> statistical techniques for empirical model building and
> model exploitation. By careful design and analysis of
> experiments, it seeks to relate a *response,* or *output*
> variable to the levels of a number of *predictors,* or
> *output* variables [i.e., factors], that affect it. The
> variables studied will depend on the specific field of
> application.

As shown in Figure 11-6, the methodology is iterative.
Each cube in the figure represents a set of experiments.
The first set determines crude contours of the process

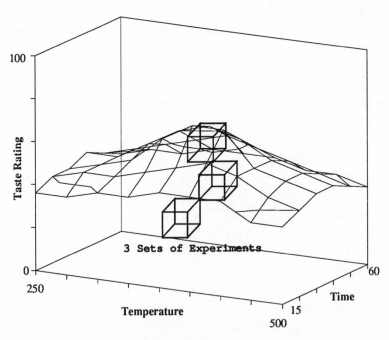

Figure 11-6. Transversal through the process space.

space by identifying which factors included in the experiments are the most significant. This tells you the direction of the local optimum. Then, in the next set of experiments, you can change the values of the factors (or the factors themselves) to get a more precise understanding of the process space.

As Box and Draper point out:

> ... any group of experimental runs should be aimed at the overall furthering of knowledge rather than just the success of any individual group of trials. Our problem is to so organize matters that we are likely in due course to be led to the right conclusions even though our initial choice of the region of interest, the metrics, the transformations, and levels of the input variables may not all be good. Our strategy must be such as will allow any poor initial choices to be rectified as we proceed. Obviously, the path to success is not unique. ... It is not the *uniqueness* of the path that we should try to achieve, therefore, but rather the probable and rapid convergence of an iterative sequence to the right conclusions.

> This iterative process of learning by experience can be roughly formalized. It consists essentially of the successive and repeated use of the sequence conjecture–design–experiment–analysis.

HOW TO PERFORM THE EXPERIMENTS

Fisher had problems performing experiments with crops because they took a long time, and it was difficult to repeat experiments under similar conditions. Most business environments pose the same problems. If an order fulfillment process takes sixty days, it will be difficult to run multiple trials of different experiments.

By using computer simulation, the process can be executed in milliseconds instead of days or weeks. In the "hunt and gather" phase, we introduced the use of computer simulation to help quantify the performance of the current process. As you are evaluating design alternatives, you can modify the models you built in the earlier phase to include the changes you are considering making to one or more factors in the process. The experiments, then, actually are carried out by executing different process models. Using factorial designs, you can measure the effects of individual elements of your new design as well as the effect of changing multiple elements. An example of this interplay between experimentation and simulation is provided in Chapter 15.

12

Key Concepts in Information Technology

One of the hallmarks of reengineering is how technology is used to enable a redesign that otherwise would not be feasible. In *Reengineering the Corporation*, for example, Michael Hammer and James Champy provide examples of how technologies such as mobile computing and expert systems can allow a company to operate in entirely new ways.

However, introducing new technologies is itself not an answer to most business problems. Only after the problem domain has been carefully studied can an appropriate technology be selected and applied. An example of misapplying technology is provided by a suggestion of Thomas Edison. He proposed a technology for mechanizing the voting procedures for the U.S. Congress. Using his technology, each congressman would have a panel (labeled "Yea" and "Nay") and would press the appropriate button whenever a bill was put to a vote. The vote could then be tallied in seconds rather than the hours it normally took to manually count slips of paper with each vote. Some months after his proposal, a test was conducted. The results proved disastrous: though the machines worked just as Edison had promised, they did not address the root causes of the apparent problem. The time it took to count the votes was actually a necessary part of the process. That time was devoted to negotiating and debating. Edison's mechanization addressed a problem that did not exist, and his proposal was rejected.

Thus, instead of addressing the entire range of new technologies that could enable new business processes, we describe in this chapter four technologies that enable reengineering itself. These four technologies—distributed computing platforms, client-server architectures, work flow software, and application development tools—together facilitate the creation of new designs and enable the iterative development of those designs.

DISTRIBUTED COMPUTING PLATFORMS

As the computing needs of businesses have grown, so have the machines that support those needs, culminating in the monolithic computers of the mainframe era. However, there are inherent problems with the use of mainframes. First, they are too expensive. The cost of computing power has fallen precipitously, but the price of mainframes has not. Second, buying or leasing a multimillion-dollar machine means a long-term commitment to a single vendor. However, the rapid changes in technology and in computing needs do not lend themselves to such commitments. What do you do with your three-year mainframe lease (or worse, your very *own* mainframe) if your company downsizes and reduces its computing needs by 50 percent? What if you want to take advantage of a new technology not available in a mainframe environment?

For these reasons, there has been a shift away from centralized computing and toward distributed computing. A distributed computing system is defined as "one in which multiple autonomous processors, possibly of different kinds, are inter-connected by a communication subnet to interact in a co-operative way to achieve an overall goal."[1] That is, instead of using a monolithic mainframe to support all of a business's information processing (an "overall goal"), the information system is comprised of a number of computers connected by a network and running special software so that the work of the processors can be coordinated. Currently, smaller computers have a superior price/performance ratio. Combined with fiber networks and distributed messaging software, distributed sets of small machines are attractive alternatives to mainframes. In general, distributed systems are designed to have a single system image (appear as one system to any user), achieve better performance, and provide higher reliability, all for less money than a single processor system.

The telephone system is a canonical distributed system, which consists of a wide range of distinct processors with multiple functions. These processors are spread over a wide area, are connected by a high-speed network, and employ a sophisticated signaling protocol to ensure coordinated processing even in the case of failures. Perhaps most important, all of these details are hidden from the user. Instead of being aware of the many components, their individual functions, and the coordination being

---

[1]Ananda, A. L. and B. Srinivasan (eds.) 1991. *Distributed Computing Systems: Concepts and Structures,* IEEE Computer Society Press.

effected, the user perceives the telephone system as a single entity with a simple overall function.

Telephone systems demonstrate the major benefits of distributed systems: low cost, fault tolerance, and scalability. The low cost is due to the reliance on many smaller machines rather than one or two large machines, as evidenced above. If everyone's telephone were connected to a single switching machine somewhere in the middle of the United States, it would cost more than pennies per minute to call across the country.

Fault tolerance means that if a processor fails or if the network between some processors fails, then the system still can function in some meaningful way. Like any system, distributed computing systems experience both hardware and software failures. For hardware failure, the typical strategy is to incorporate redundant processing components and provide special software to manage these components in time of failure (e.g., by recognizing a failure state and switching to the reserve components) in order to mask the failure from the rest of the system. Masking a failure means that the system continues to provide the standard specified service despite a component failure. For example, a system could mask a disk failure by providing a disk dually ported to different storage controllers. Then, if the controller failed, the operating system could detect the failure but mask it from the rest of the system by switching to the redundant controller. The telephone network has several (redundant) routes to complete a call between two points; so if a switch fails, the signaling software detects the failure and selects another route not including that switch.

Last, distributed systems are scalable because you can add processors as you need additional capacity. Moreover, these processors do not all have to be the same. Different types of processors can communicate because interface standards have been established, and each vendor adheres to the standards. As more people use the telephone networks, there is no need to replace all of the machines. Rather, new machines are added to the network, and the calling routes are redistributed. In the United States, the technology ranges from rotary phones and mechanical switches to multiprocessor digital switches complete with expert systems for fault diagnosis.

Besides flexibility, an important advantage of having heterogeneous machines communicate is that they can share data. Consider the college registration process shown in Figure 12-1. Once a student completes a registration form, the three-step process consists of checking the student's status in the student profile database; then, checking the scheduling and availability of courses, and, if everything is okay, updating the course lists; and, finally, sending the necessary data (e.g., student ID, course credits, etc.) to the billing system.

Figure 12-2 shows the systems architecture supporting the process. In this example, these systems—student profile, course scheduling, and billing—were each developed at different times by different organizations. The systems supporting each step in the process do not share data but rather communicate via simple interfaces. Notice that some of the student profile data is needed by each of the three systems, and so the data is sent every night from the student profile system to the other two systems in an

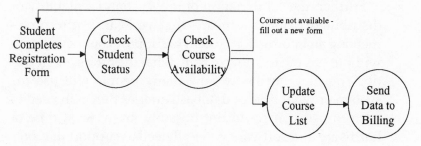

Figure 12-1. A college registration process.

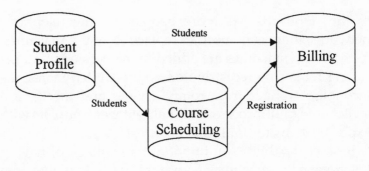

Figure 12-2. The systems architecture supporting college registration.

attempt to synchronize the redundant versions of the same data.

Distributed platforms enable the reengineering of processes such as the college registration process by facilitating data sharing and iterative development. With software that allows you to communicate across systems and maintain a single system image, you can gradually eliminate the type of data redundancy shown in Figure 12-2. Only one system should update the student profile or registration data; the other two systems could read directly from the same physical copy of the database (Figure 12-3). This eliminates the need for a synchronization mechanism and, more important, eliminates the problems that arise from inconsistencies between the different copies of the data.

Further, the combination of low cost and scalability of distributed platforms means that you can begin implementing an information system to support your new design without having to know exactly what the new system will do. You may not know how many people will use the system or what type of database queries they will need to perform; so you could not possibly know what type of hardware and software you will need to support the complete design. Thus, distributed platforms lend themselves

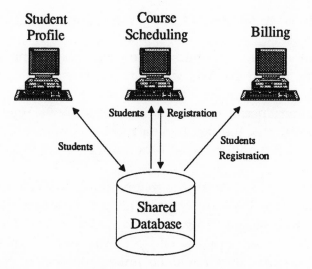

Figure 12-3. The distributed systems architecture.

to an iterative development approach and let you minimize the risk associated with buying or leasing expensive
equipment or making premature system design decisions.

CLIENT-SERVER ARCHITECTURES

Imagine that you had to give someone instructions to mix
the perfect frozen margarita.[2] You could certainly provide
all of the necessary steps (or "logic") on a single napkin or
sheet of paper. Now, imagine that you had to write down
the instructions for building the space shuttle. You might
need millions of sheets of paper, and organizing the sheets
would be next to impossible. "Check the structural integrity of the O-rings. If the seal is not disturbed, proceed
with tightening the bolts (part 311-RB). If the seal is not

---

[2]The recipe is included in the appendix.

intact, go to section 18A.2 for instructions on tightening the gaskets."

In the past, building software was like giving space shuttle directions. You would write down all of the instructions in sequence and include logic ("GOTO's") for handling special cases. As programs grew larger and more complex, the logic became more and more enmeshed. The result would be what programmers refer to as "spaghetti code."

The way to untangle the spaghetti was to create modular software. Instead of putting all of the instructions and logic in a single program, programmers would create smaller modules that would perform specific functions. The modules then could be linked together to form a single application. When a user or an analyst needed a different application, the programmers first would determine the functions that were required, and then they could reuse the modules they previously had created.

Client-server architectures advance this concept even further; they divide programs into two general categories. A *client* is a program that initiates a request for some resource, such as a computation or data. A *server* receives a client's request, performs a function (such as performing the calculation or database query), and returns the result. Clients do not know how servers are carrying out their functions. The only thing that clients know about servers is how to request their resource. In general, the physical implementation of individual clients and servers is abstracted from the rest of the system. Thus, a database server may be rewritten to interface with a new database management system, but the rest of the system is unaffected as long as the means of requesting the resource remains the same.

A modular architecture that promotes software reuse and provides robustness by abstracting the individual components of the system will facilitate the evolutionary development of an information system supporting your new business design. As you iteratively implement new design ideas, you can create (or redesign) clients and servers without having to change the rest of the system.

WORK FLOW SOFTWARE

Across the reengineering projects we have been involved with, one common root cause of process problems has been that system functions are not tied to the tasks in a process. There are four common symptoms of this root cause:

1.   *A lack of system support for tasks in a process:* As there are changes to the process, there are no corresponding changes to the supporting information systems. This results in "work-arounds": manual tasks and tasks whose only purpose may be to satisfy a vestigial system constraint.

2.   *An inability to track the status of an activity:* If the tasks are not explicitly related to system functions, there is no way automatically to track the progress of an activity. This results in ad hoc tracking mechanisms, ranging from elaborate folder designs to information systems dedicated to tracking. In either case, the status information is of poor quality because status is recorded independent of the actual execution of a task.

3.   *An inability to provide realistic due dates to customers:* To provide a date when work will be performed, you must know how long it will take to do the work, and you must know how much work you already have to do. Knowing these things requires access to data that will affect the work (e.g., customer site data that might affect installation) and requires data about the current workload (e.g., that all the installers already are booked for the next two weeks). Without these data, most companies resort to quoting "standard" intervals ("We'll have someone there in five business days, sometime between 8 A.M. and 5 P.M."), not knowing whether the company can meet the commitment it made to the customer.

4. *The creation of independent mechanisms for compiling process metrics:* To compile statistics on cycle time, throughput, and so on, you need somehow to track when tasks begin and when they are completed. If the existing support system functions are not tied to tasks in the process, then the metrics must be compiled manually, usually with the aid of a quality measurement system. As in the above case, the process metrics will be inaccurate because the measurements have been made independent of the actual execution of the process.

One way to tie system functions to tasks in the process (and to eliminate the above symptoms) is to employ work flow software[3] on a client-server architecture. Work flow software contains an on-line definition of a business process and guides the progress of an activity as each task is executed. (In an order fulfillment process, for example, the work flow software could guide orders through the process.)

Work flow software allows a user to define a process on-line, including the dependencies between tasks and the system functions that are tied to specific tasks. In a client-server architecture, a system function would be a service that is explicitly related to the execution of a task in the process. When all of a task's predecessors have been successfully completed, the service associated with that

[3]We have seen both custom-built and commercially available work flow software. Commercially available products include ProcessIT from NCR, InConcert from Xerox, and FlowPath from Honeywell Bull.

task then is executed by the work flow software. The task could be automated or manual. Thus, the service associated with the task could present a screen to a user, could execute a database query, or could exchange data with a remote system.

In addition to providing facilities for defining a process, most work flow software provides at least four other basic functions: work assignment and routing, scheduling, work list management, and automatic status and process metrics.

Suppose you are in charge of a business that delivers floral arrangements nationwide, and your delivery process is supported by a new information system, complete with work flow software. The process (shown in Figure 12-4) begins when a customer calls an 800 number for an arrangement. A sales representative uses the information system to create a P.O. ("petal order") and thereby triggers the floral arrangement delivery process. Before using the work flow software, the sales reps (all located in Podunk, Iowa) would use a job aid to find out which local florist was closest to the delivery address. Then, a sales rep would print out the P.O. and fax it to the local florist. When the local florist completed the order, it would call

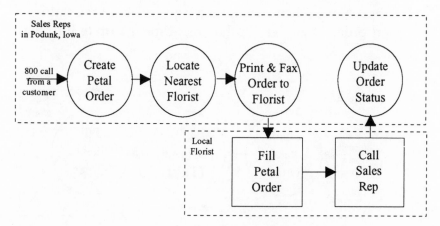

Figure 12-4. The petal order fulfillment process.

back with a status update, and the sales rep would update the order with that information.

In their new environment, shown in Figure 12-5, the florists have access to the same system as the sales reps. After all the necessary order data are entered into the system, the work flow software assigns the next task in the process to the nearest florist. Because the assignment task is automated, it also can be made more powerful. For example, it can perform load balancing in a given area, ensuring that one florist does not get a hundred orders while another local florist gets none. The work flow software also can assign weights to each florist according to past performance, thus routing more work to florists who have performed better in the past.

Each florist has a "work-to-do list" on its terminal. As the work flow software routes work, the work lists are updated automatically. Printing and faxing of orders are eliminated as a means of communicating assignments. Florists check their work lists, and then execute the next task in the process (which might be reviewing inventory or dispatching a worker to make a delivery). As the florists manage their work lists, the status of each P.O. is updated automatically and made available to the sales reps. Now, when a customer inquires about the status of an order, the sales rep has real-time, accurate informa-

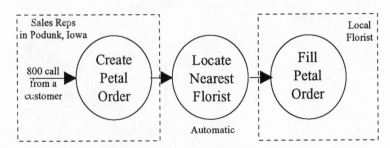

Figure 12-5. Process streamlining enabled by work flow software.

tion at his or her fingertips. Similarly, process metrics are calculated automatically by summing execution times of tasks.

By using work flow software and providing all people involved in the process with access to the system hosting that software, companies like the national floral delivery service can enable a new process design (or even achieve significant improvements in their existing processes).

With mature commercial products, businesses can use work flow software to surround the existing information architecture. (For example, most products already have means of interacting with existing databases and executing different applications on different machines.) Thus, some of the current systems problems can be abstracted from the case teams, and new processes can be modeled quickly. This is essential for testing process changes and for realizing benefits rapidly.

APPLICATION DEVELOPMENT TOOLS

Ever since the development of computers and computer languages, technologists have been looking for easier ways to specify what a computer should do and how it should do it. Figure 12-6 shows the evolution of a code fragment

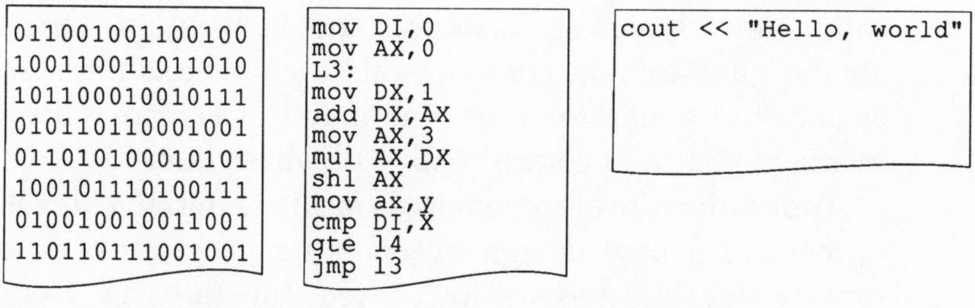

```
011001001100100
100110011011010
101100010010111
010110110001001
011010100010101
100101110100111
010010010011001
110110111001001
```

```
mov DI,0
mov AX,0
L3:
mov DX,1
add DX,AX
mov AX,3
mul AX,DX
shl AX
mov ax,y
cmp DI,X
gte 14
jmp 13
```

```
cout << "Hello, world"
```

Figure 12-6. The evolution of software specification.

that prints a string of text on a screen. Forty years ago, programmers specified the exact sequence of zeroes and ones code that the computer would process. Some ten years later, computer scientists developed software that took as input a higher-level specification (assembly language) and translated it into the bits needed by a computer. Shortly thereafter came the advent of interpreters and compilers that translated still higher-level languages (e.g., FORTRAN, C) into the necessary zeroes and ones. Each new means of specifying software functions abstracted unnecessary details (such as the number of internal registers) from the programmer.

These advances greatly improved programmer productivity. However, just as there were advances in programming specifications, so were there advances in hardware and operating systems and user interface mechanisms. Thus, even with high-level programming languages such as C, the job of specifying what a computer should do and how it should do it remains a complicated task.

One way around this is to create yet a higher-level specification of software that abstracts details of the hardware and software platform from the programmer, leaving the programmer with only the functions to specify. Code generators translate a higher-level specification into a lower-level language, such as C, which then is compiled into zeroes and ones. Typically, code generators are designed to generate software for a particular function, such as presenting text and graphics on the screen. Application development tools do this and, in addition, generate the software that links several functions together. An application generator thus might link a screen with a message-passing protocol with a database query.

Application development tools facilitate iterative development of a new design by enabling programmers to rapidly develop and deploy system functions to users.

Because most of the code is generated,[4] changes can be made quickly. Changes to screens or queries may require only modifying the graphic image of the screen or else a few lines of the high-level specification. The thousands of lines of source code (e.g., C language) needed for those screens and database queries, as well as the code for communications and other functions, then can be regenerated in minutes.

Suppose your operations personnel used different workstations (e.g., personal computers, character-based terminals) to access the company's information systems. To develop a different user interface for each workstation and to maintain multiple sets of source code would be costly even if your systems development organization had the necessary skills. Currently, however, code generators are available[5] that generate user interface software for different workstations based on a single, high-level specification. The intricacies of designing screens on different workstations are abstracted from the programmers. Thus, they can quickly demonstrate the look and the feel of a new screen or make changes to an existing screen.

Further, because the high-level specification files do not include details of the hardware or software platform, you can change the platforms without changing all of the specifications. You simply change the relevant generator and regenerate all of the source code. Thus, if you make a mistake in designing a new system, or if you need to add another component to the platform, the impact of the design change is limited to a single piece of software—the generator. For example, if you need to switch database

---

[4]Typical expansion rates are in the thousands. That is, for every one line in the high-level specification there are thousands of lines of C code.
[5]Commercially available products such as XVT and JAM generate code for multiple user interface platforms.

vendors, you will not have hand-crafted, vendor-specific code strewn throughout the system. Instead, you just need to modify the tool that generates the database accessors.

Application development tools extend this concept of abstracting implementation details from the program-mers and provide even greater benefits in terms of quickly delivering system functions to users.

SUMMARY

The four technologies described in this chapter allow for developing applications more quickly, more cheaply, and with better quality than ever before possible. Yet, these technologies are not sufficient for all problems. For example, better ways of distributing information (e.g., hand-held computers) and of interacting with machines (e.g., voice recognition) are now emerging. However, even the new-est technologies will face problems in being integrated into business environments. Therefore, the ideas described in this chapter will be essential for successfully reengi-neering businesses using both current and newly emerg-ing technologies.

13

Data Analysis and Reengineering Projects

This chapter describes a number of ways to identify the data required in a process, and what you need to do with the data once you have identified them. As stated in Chapter 10 and throughout this chapter, we do not advocate a protracted analysis period using CASE tools (or the like) prior to delivering system functions to the case teams. This statement runs counter to much of the literature on reengineering, yet our experience is that analysis methods and tools too often become ends in and of themselves. Moreover, protracted analysis at the front of the project is obviated by the use of an evolutionary approach.

Given that reengineering projects are started because of a dire business need, a time-consuming, detailed analysis usually is not appropriate. Thus, unless your project has the resources to make a long-term commitment to ongoing analysis methods and to maintaining requirements using those methods, you should avoid a complicated analysis methodology.

The sections below provide an introduction to three analysis methods—logical data modeling, structured analysis, and object-oriented analysis—as well as automated support of these methods. In reengineering projects (and in most software development projects), methodologies are like Robert Louis Stevenson's "bright face of danger." They are attractive insofar as they provide consistency and direction, and they are useful when the environment is stable and there is already expertise in applying a methodology. However, the search for methodology has been the death knell for many a project. Too often the quest for defining how to do the work overtakes actually doing the work, and the project goes nowhere.

METHODS OVERVIEW

This section identifies three, somewhat overlapping, methods: logical data modeling, structured analysis, and object-oriented analysis. The goals of using any of these methods in a reengineering project are the same, no matter which methods are used:

- To document the data used in the process.
- To communicate which concepts are fundamental to the business and how they relate to each other.
- To serve as the foundation of database applications that will support the new process.

You do not need to be a computer scientist to apply these methods. All it takes is some common sense and some organizational skills. Later in this chapter we will describe how to judiciously apply these methods to avoid the danger described above. We will also describe specific tools that you will need to make this part of the project successful.

Logical Data Modeling

An excellent guide to data modeling is the *Handbook of Relational Database Design* by Candace C. Fleming and Barbara von Halle. They define logical data modeling as "a technique for clearly representing business information structures and rules as input to the database design process."

Structured Analysis

In his text on structured analysis, Edward Yourdon asserted that it was "more interesting than anything I know, with the possible exception of sex and certain types of Australian wine." Well, it is not quite that interesting—but it can be practical if you need to understand the data and the logic in a business process.

Structured analysis aims to produce three models or views of a business process: a data-flow diagram for modeling functions, an entity-relationship diagram for modeling data, and a state transition diagram for modeling changes over time. These are three views of the same business, and there are rules (see Yourdon's *Modern Structured Analysis*) for "balancing" the models—that is, checking for consistency across them. The models are meant to be technology-independent; so the requirements (logic and data) should be immune to changes in technology.

The point is to model the data in a process, not in a database application.

Object-Oriented Analysis

As noted by Booch in *Object-Oriented Design*, the term object-oriented "has been bandied about with carefree abandon with much the same reverence accorded 'motherhood' [and] 'apple pie'." It is chic to say that an information system is object-oriented, but the term has a very specific meaning. Whereas structured analysis focuses on the flow of data within a process, object-oriented analysis focuses on objects and their behaviors. The abstraction of the real world is done in a very different way here. Instead of thinking in terms of a process transforming data, one identifies real-world objects (e.g., a student or course list) that act as "autonomous agents that collaborate to perform some higher-level behavior."

Given these characteristics of object-oriented modeling, there are great advantages to using object-oriented analysis. However, in our experience it has not been intuitive for business planners and operations personnel to think of their business in terms of objects and their behaviors. They have felt most comfortable in talking about the process. (It took W. Edwards Deming fifty years to get businesses to think in terms of processes instead of isolated events. It should not be surprising for it to take at least a while for businesses to think in terms of objects.)

THINGS TO WATCH OUT FOR

Although data analysis methods are necessary for understanding the data required by the process, a number of problems with them should raise a big red flag. If you encounter any of these problems on your project, it may well be a sign that your analysis has gone out of control and that you should scale the analysis down in some way, depending on the problem. Each section below describes a specific problem and corresponding corrective actions.

Prolonged Ramp-up Period

Methods complexity and training requirements tend to delay the date when productive work can begin. Although training is important to ensure the quality of the analysis methods, needing more than a few weeks of training is a sign that the methods are too complicated or contain too much new information. If a month after the beginning of the "innovate and build" phase people still are confused about the methods and their role in the project, consider introducing fewer new concepts. For example, a project might use structured analysis and logical data modeling and have those methods drive the database development, and might have all project members (analysts, developers, and business planners) use the same CASE tool to view the data related to the process. Well, this sounds great, but if after a month no one has a clue about how this arrangement is going to work, then try introducing just data modeling or just structured analysis with limited tool support.

Battles over Tools and Methods

Surprisingly, the choice of specific CASE tools and methods sometimes has caused project-endangering political rifts. Do not let this occur. If there is a disagreement over tools or methods, select a group to perform the research and make the decision. Make sure that everyone on the project understands how the decision will be made, and that once it has been made the project must move on. The continual revisiting of a past decision on CASE tools has introduced unnecessary churn and confusion on a number of projects.

Increased Dependence on Consultants

Of all the red flags, this is the biggest. Bringing in consultants at the early stages of a project is fine for planning. However, if you need more and more consulting time for data model reviews and methodology instruction, then your project is in over its head in terms of analysis methods. In one project, after weeks of training and months of analysis, a consultant was hired to make sense of the mounds of data-flow diagrams that were being produced, which no one could use to build an information system.

If the project has an increasing need for methods consultants, consider reducing the complexity of the methods or reducing the number of new concepts as described above.

Rising Data Analysis Costs

Training, CASE tools, and consultant fees (in that order) are the main culprits in driving up analysis costs. Training in structured or object-oriented analysis typically costs $1,500 to $2,500 per person for a one-week course, and data modeling takes an additional week. CASE tool software licenses range from $500 to $10,000 per user, not including hardware or network connectivity. Finally, consultant fees may be $1,000 per day. If you do not have experience in applying analysis methods, then paying for that lack of experience will be prohibitive and will introduce additional risk.

USING CASE TOOLS

CASE tools combine graphics drawing software with database technology to handle the creation, update, and reporting of all data analysis products. There are three main benefits of using CASE tools, if they are used properly:

1. They can greatly increase the maintainability of the data and logic requirements of the business process; having the integrated requirements on-line makes it easier to estimate impacts and make changes.

2. They help to integrate the different views of the business process; the data and the logic used to carry out the business process can be viewed together.

3. They present data and processes in a leveled way; one can "step down" from a high-level view of a process to a lower and lower level of detail about the process while always having access to a similarly leveled view of the data.

By having the data and the process in one place, CASE tools make it easier to verify that the data and logic requirements "match." Such confirmation is called "balancing the models" in structured analysis. This makes it easy to check the impacts of a proposed change in the data, or to see whether it would be straightforward to place additional logic requirements on the process. However, to realize any of these benefits the project members must continuously apply the analysis methods (updating data-flow diagrams and entity-relationship diagrams) as the business requirements change.

In general, CASE tools will greatly facilitate the work of experienced data analysts, but for most projects, the tools are too costly and too difficult to administer to encourage maintenance of the models. Once the models are not maintained, the tool is nothing but a file cabinet of out-dated requirements.

Another point is that tools will facilitate creating the models but will not necessarily improve the quality of the analysis. As Grady Booch pointed out in his seminal work on object-oriented methods, "Great designs come from great designers, not great tools. . . . one of the things that tools can do is to help bad designers create ghastly designs much more quickly than they ever could in the past." The tools are good at checking for completeness and for consistency across models but not at determining whether a data-flow diagram should be further decomposed or whether the entities that have been modeled are too abstract.

14

Redesign Principles

There are no formulas for design or for redesign, but there are some principles to guide you and improve your chances for creating something worthwhile. This chapter, which is meant mainly for reference use, is intended to guide you as you iterate through the "innovate and build" phase.

As you generate new designs, use the principles in this chapter as a check. Does your design have some of the characteristics described in the "Process Design" section? Do your system requirements avoid some of the pitfalls described under "Information System Details"? Of course, the principles in this chapter may not all apply to your business design. However, if you contradict one of these principles, make sure that you have strong reasons for doing so and that you avoid the problems usually associated with not following it.

For ease of reference, the principles are grouped into these categories:

- General principles
- Process design
- Organizational structure
- Interfaces
- Automation
- Information system details

GENERAL PRINCIPLES

Keep It Simple

In *The Iliad,* Odysseus often was too clever for his own good. He outwitted the Cyclops and managed to endure the Sirens' call, but he was rewarded with ten years of wandering on the open seas. Do not let this happen to you. Just because there are some advanced tools and techniques that can support your redesigned business, do not eschew the simple tools you can use or the simple changes you can make. How many businesses have overlooked Ishikawa's seven tools in favor of costly quality management systems? Hubris often makes people introduce complicated processes and technologies into the redesign (perhaps as a way of demonstrating the cleverness of the designer). However, the goals of reengineering are to make businesses less costly and more flexible, and the simplicity of the new design is a key to achieving these goals.

Push Work Up, Not Down

"The only way to motivate employees is to give them challenging work in which they can assume responsibility."[1] Despite the evidence supporting the need to provide workers with interesting, challenging work, there is still a trend away from this. Too often, we have heard managers strive to "push work down to the lowest level" as if their goal were to employ a workforce of unskilled clerks. However, the business environment has changed dramatically since Ford put poorly trained workers on assembly lines and was able to produce cars. Today, having people perform repetitive, mundane tasks leads to poor morale and an inability to handle business dynamics. People who are trained and allowed to think are more valuable (despite their higher pay) than a business version of toll collectors mindlessly carrying out simple tasks. How much more valuable are sales representatives who know the company's products and can suggest alternatives in a pleasant manner than representatives who merely type in data as prompted by the screen in front of them? How many more products will a personable, well-trained representative sell? How many more referrals will the company get as a result of a dialogue with that representative?

In *Thriving on Chaos*, Tom Peters performs a simple calculation based on his own company's dealings with Federal Express. Peters shows that a single courier who interacts with the customer can represent $14 million over ten years (e.g., $1,500 per month for a loyal business customer times two loyal customers (one more by word-of-mouth) times 40 customers seen per day). After considering those numbers, Peters recommends that you think more about how sales and service people should be trained, how they should be supported, how their uniforms should look, and so on. A company that allows only its managers to think will not endure. Pushing work up to higher levels

---

[1]Herzberg, Frederick, One more time: How do you motivate employees? *Harvard Business Review*, January–February 1968.

provides a company with an entire workforce that can participate in the betterment of the company.

PROCESS DESIGN

Design for the 80% Case with Exception Handling for the Tougher Cases

In his seminars, Michael Hammer points out that most processes are "overengineered," that is, designed for the worst case. In our experience, analysts, systems developers, and operations staff are fascinated with the rare cases. Interview sessions often are cluttered with anecdotes and "what-if" scenarios that are interesting but do not help in designing a simple process. A manager might remark, "What if the customer has no credit, no address, and marks his name with an 'X' but the supervisor says it's okay? That happened last year, you know," and then follow up his remark with a lengthy recounting of the once-in-a-lifetime scenario that the new design should handle.

In one example, a telecommunications support system was delayed because analysts were trying to figure out how to designate rows and slips in a yacht basin using the address fields built into the system. Even though services to yacht basins represented less that .05 percent of the total business, the "rows and slips problem" prevented the company from getting the benefit of employing the new design for the more common scenarios.

The 80/20 rule applies here. You will get 80 percent of the benefit from 20 percent of the effort. By designing for the common case, you derive that benefit *at the beginning* of the reengineering project. Instead of creating a robust but complex process to handle all situations, you should aim to create a simple process that handles the simple situations. Then, even if you must resort to manual, labor-intensive steps to handle the more complex cases, you only will be employing those steps a small percentage of the time.

Validate Data at the Source

If you wait until the end of a process to check the validity of data, you still will have bad data, but you will have also spent a significant amount of your resources to promulgate that bad data through the process. Whether it is a customer order form or an invoice from a vendor, do not wait until the end of your process to check for errors. In one case, a billing system would reject orders due to data errors weeks after the order originally was placed. What is the chance of finding the cause of an error after so much time? How many resources will be needed to deal with the consequences of such errors? To eliminate the need for Byzantine error-handling procedures, your new design should deal with the consequences of errors when they are introduced into the process.

Eliminate "Review" Tasks

Performing a task whose only purpose is to review the work of someone upstream in the process is akin to waiting until a finished product is available before inspecting for quality. As Deming and others have strongly protested, mass inspection is not the way to improve quality; instead quality should be *designed into the product*. Similarly, tasks should be designed only to produce high-quality output, whether it be data or product. Relying on later tasks to review the work only will result in higher costs and will engender suspicion between the inspectors and those being inspected while not producing any significant improvement in quality.

ORGANIZATIONAL STRUCTURE

Listed below are principles for organizing personnel to execute the business process. Some of these topics also are discussed in Chapter 16.

Organize Work Groups around Extended Processes, Not Tasks

There are three problems with organizing work groups by task or by function:

1. Each group tries to optimize the performance of its own function.

2. Each group tries to procure its own tools to support its specific needs.

3. Each group insists on inspecting the work of other groups to ensure that its own quality measurements are not affected.

This results in a fragmented process. Over time, the different work groups become more and more like separate fiefdoms, complete with local lords and interfiefdom battles. As each group focuses on its own performance relative to the other groups, all of the groups forget about the customer.

Organizing work groups around extended processes helps each individual to better understand how his or her own work relates to a product or a service the company provides. As you develop teams to execute the new design, ensure that each team's role is clearly defined in relation to an extended process.

Involve As Few People As Possible in the Performance of a Process

When Adam Smith described the benefits of functional organizations, he used for his example workers in a pin-making factory. Instead of having each worker manufacture complete pins, Adam Smith assigned each worker a simple task, such as cutting the proper length of wire, and the worker was to perform the task over and over again. This division of labor boosted productivity. Further, adding more people to the process resulted in the generation of more products. However, things have changed in the last 200 years. In the classic software engineering text *The Mythical Man-Month,* Frederick P. Brooks summarizes why involving more people in a complex process (such as system development) may not result in extra production:

> Men and months are interchangeable commodities only when a task can be partitioned among many workers *with no communication among them.* This is true of reaping wheat or picking cotton; it is not even approximately true of systems programming. . . . In tasks that can be partitioned but which require communication among the sub-tasks, the effort of communication must be added to the amount of work to be done. Therefore the best that can be done is somewhat poorer than an even trade of men for months. . . . If each part of the task must be separately coordinated with each other part, the effort increases as $n(n-1)/2$. Three workers require three times as much pairwise intercommunication as two; four

require six times as much as two. If, moreover, there need to be conferences among three, four, etc., workers to resolve things jointly, matters get worse yet. The added effort of communicating may fully counteract the division of the original task. . . . Since software construction is inherently a systems effort—an exercise in complex interrelationships—communication effort is great, and it quickly dominates the decrease in individual task time brought about by partitioning. Adding more men then lengthens, not shortens, the schedule. (pp. 17–19)

Using small teams (no more than eight to ten people) to execute extended processes will minimize the communications overhead while still providing good productivity per team. (Ideally, try to use one person, and then add more as needed.)

Use Coaches instead of Supervisors

Teams that understand their roles in the company and how to play those roles do not need supervisors. Supervisors check the status of things, sign timecards, and take classes in leadership. Instead, a good team needs a coach—someone with an in-depth knowledge of the business—to help make the team better. When the team is faced with a problem it has not dealt with before, the coach works closely with it to transfer knowledge and experience to it.

INTERFACES

Simplify the Customer Interface

The interaction between the customer and the company is the area where most reengineering breakthroughs are made. Too often, the blame for process errors is placed with the customer: "If those damned customers would only fill out the form correctly, things would go more smoothly." The "it's the customer's fault" syndrome is a symptom of a poorly designed interface between the customer and the company. Lengthy forms are typical examples of poor interfaces, as are forms asking for information that the customer may not readily know. Think of credit applications. In addition to the personal data they require, they ask you for account numbers and account balances. Yet, how many of us know all of our credit card account numbers when we apply for a loan or for additional credit? Moreover, why should we be asked for such detailed information if the lender typically checks the information with a credit bureau anyway?

Whenever you are designing a customer interface, be aware of how you feel when you are the customer. The act of communicating your need for a product or a service should be simple and quick. Confusion about how to order or a delay in taking an order bewilders the customer and loses sales.

Reduce Dependence on E-Mail, Fax, and Telephone Communication among Work Groups

Although electronic mail, fax machines, and telephones have become essential tools in conducting business, they are not designed for coordinating work or sharing structured information among groups. All three media are subject to delays in delivering information, whether the delay be the result of transmission (e-mail), delivery (fax), or voice mail (telephone). (Frustrated by trying to reach people by telephone, a colleague of ours performed an unscientific sample to determine the average delay in phone communication and found that upwards of 90% of people in one work center had their calls automatically routed to voice mail.) Instead of trying to tie groups together using these technologies, endeavor to bring functions together into co-located teams. Where that is not feasible, use information systems and technologies such as work flow software (described in the Chapter 12) to execute a process with multiple work groups.

AUTOMATION

Avoid Automation for Its Own Sake

More often than not, automation is not a panacea but a plague. In an effort to reduce costs and improve quality, General Motors spent billions on deploying robots in its factories. Yet, because of problems with the robots and a failure to understand some of the real root causes of its quality problems, GM failed to achieve any of its lofty objectives. Also, merely applying automation to an existing process is like "paving the cow paths."[2] The symptoms may change, but the root causes will remain. Implementing high-technology initiatives involves risk because of the cost and the complexity, and you should be sure that the introduction of new technologies will address the root causes of current problems and enable the development of a new way of doing business. You can avoid ineffective applications of technology by having a process orientation in building systems and by using an evolutionary development approach to test ideas with maximum flexibility.

---

[2]Hammer, Michael, Reengineering work: Don't automate, obliterate, *Harvard Business Review*, July–August 1990.

Mechanize Routine, Mundane Tasks before Mechanizing Interesting Work

When expert systems were introduced, there was a clamor for software that could do the work of highly-skilled professionals. Knowledge engineers worked with the best, most experienced staff to encode their understanding of the business, but administrative assistants working in the same process might still be hand-carrying boxes of paper to deliver work from one group to another. Why should we automate the most interesting part of the process if there is a great deal of mundane work that should be eliminated or automated? People like to do interesting work; they take pride in using their knowledge and experience to make decisions, whether in customer service or in circuit design. Instead of robbing people of that pride, focus your efforts on eliminating such tasks as the re-keying of the same data into multiple systems, the checking of other people's work, or the print–fax–file cycle.

INFORMATION SYSTEM DETAILS

Beware of "Information As a Corporate Resource"

Recently, crusades have been waged under the banner of using "information as a corporate resource." Is it important to provide people with the information they need to make decisions? Of course. What must be avoided is the broadcasting of information to every computer in the corporation so that managers can make work for their staffs and for themselves. As Schrage points out, ". . . as anyone who's worked for a living knows, organizations don't run on information. They run on relationships with customers and suppliers and relationships between peers and colleagues."[3] When faced with a request for access to data, be sure to operationalize the request; that is, find out exactly who will use the data, how often, and to what end.

Avoid Reverse Engineering of Existing Databases

Reviewing corporate databases is sometimes akin to archaeology, in which you can examine a specimen and see the vestiges of previous generations. If the new design simply converts the existing databases to a new platform, then you will be including all of the vestigial data in the new design. If you are streamlining an ordering process, for example, do you still want to store the date, time, initials, title, and maiden name of the person who authorized an order? With such fields in a database, there is a tendency to spawn tasks designed to provide values for those fields. Avoid the confusion associated with storage and maintenance of unneeded data.

Avoid Coding Data

In the past, when computer memory and communications bandwidth were scarce resources, efficiency was the most important requirement in representing and storing data. One way to reduce storage requirements was to store data using codes instead of the actual values—"Product No. QRZ13A" instead of "The red turtleneck, size extra large." Unfortunately, improving the efficiency of storing a data field often was accomplished at the cost of reducing the semantic value of that field. If something is important

---

[3]Schrage, Michael, *Shared Minds: The New Technologies of Collaboration.*

enough to store in the database, it is important enough to present to users in a meaningful way. Internally, algorithms and databases can use more efficient methods, but there no longer is a need to code data. If coded representations must persist, then, at a minimum, always have the meaningful identifier visible or readily accessible.

Avoid Reports More Than a Few Pages Long

In dealing with information, more is not necessarily better. As described earlier, for a time we passed a large dumpster that was filled with discarded reports *every day*. Some of these reports were hundreds of pages long, far too long for any person to make sense of them. Yet each day the reports were generated, printed, distributed, and filed or thrown away. Before agreeing to provide a report, take the time to find out what will be done with the information and whether there is a better way of presenting it than in report form.

15

The Redesign Process: A Step-by-Step Example

We now have described both the methods and the means of redesign. In this chapter, we present an extended example illustrating how the two may be interwoven into a redesign process. The first part of this chapter describes a reengineering team's presentation of its results from the "hunt and gather" phase of its project. Then, the team's redesign process is chronicled as it first generates redesign candidates and then iteratively validates elements of the new design. The sum of these iterations will be a design that is radically different from the existing environment.

In Chapter 10, we introduced a fictitious company named Comic Delivery, Inc., whose main business is to deliver jokes to customers. The processes that comprise Comic Delivery (described below) represent an amalgam of the business processes we have examined, and are common

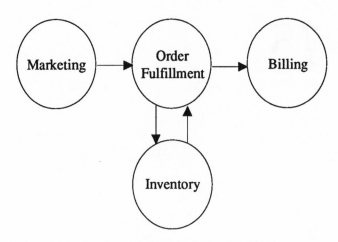

Figure 15-1. The processes of Comic Delivery, Inc.

to most businesses that deliver goods or services (see Figure 15-1). These processes are representative business processes, underscoring the generic nature of the redesign approach espoused in this book:

- *Order fulfillment process:* accept customer orders, schedule and coordinate delivery, and provide timely, accurate data to the billing system.
- *Inventory process:* continually update the different joke databases, including puns, anecdotes, and one-liners. This sometimes includes the purchase of entirely new databases.
- *Billing process:* send out a bill to the person who requested the delivery and handle inquiries about bills.
- *Marketing process:* handle all advertising and sales promotions.

Comic Delivery also has other processes (referred to as "enabling processes" by Michael Hammer) that support their main business. These range from administering human resources to providing legal services.

Comic Delivery embarked on a reengineering project because it was losing its competitive edge. At one time, the company was the only national distributor of comedy. As a monopoly, it had a constant stream of customers even though its service was not always fast or reliable. With the advent of competition, however, Comic Delivery's market share began to erode. Its top management sought radical measures to shore up its position as the market leader.

THE PROBLEMS

Everyone at Comic Delivery knew there were problems with the order fulfillment and the billing processes. Each month, the number of customer complaints was published in the company's quality newsletter, and the numbers were embarrassing. Of their 10,000 deliveries a month, more than half involved some sort of rework:

- Deliveries were sent to the wrong address.
- Customers were sent the wrong material.
- Customers were being billed incorrectly.

This rework resulted in long cycle times, high costs, and lost revenue.

After a proliferation of quality improvement teams failed to improve the situation, the management of Comic Delivery formed a reengineering team to look at the order fulfillment and billing processes. Initially, their team consisted of six people:

- Two reengineers: a reengineering consultant to help guide the project and a computer scientist with some relevant business experience.
- Two insiders: a highly-respected worker currently

supporting one of the current business functions
and an older employee who had worked in many
different areas of the company.
- A skilled public speaker to serve the public relations
needs of the project.
- A supervisor in the systems division who was
generally viewed as competent and objective and
would serve as the team's champion.

The team spent six weeks studying the operations processes and the information systems architecture, attempting to identify the root causes of the rework problems.
One of its first decisions was to change its own scope.
Based on interviews with operations personnel and examination of billing error reports, the team concluded that
the billing problems were not part of the billing process.
The generation of bills and the handling of billing inquiries actually worked well; the data on the bills were causing
the problems. The root causes of billing errors occurred
earlier in the processing of customer orders, in the order
fulfillment process.

The team built a simulation model of the existing order
fulfillment process, depicted in Figure 15-2.

The process begins when a customer calls in with a
request to send a joke (perhaps a specific joke or a joke on a
certain topic) to another person. Customer representatives take down administrative information from the
customer, such as where to deliver the joke and whom to
bill it to. Then, the representatives access multiple systems to search large databases and present the customer
with several jokes that meet the search criteria. Once the
customer makes a selection, an electronic mail message is
sent to a group that manages the actual delivery so that it
can select a courier and coordinate schedules. When things
work well, the courier appears at the appointed date and
time and tells the joke. At the end of the day, each courier

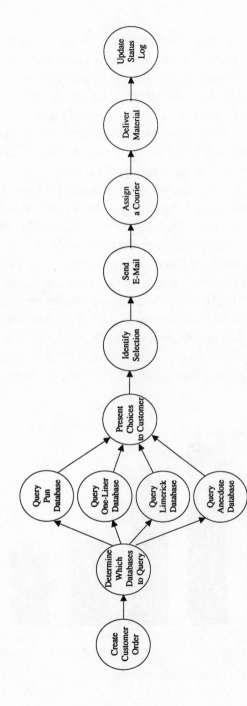

Figure 15-2. The Comic Delivery order fulfillment process.

logs in all the deliveries that were made successfully as well as the reasons for unsuccessful deliveries.

At the end of six weeks, the reengineering team delivered a presentation of its findings to senior management. The team's presentation had three main parts: describing *what* Comic Delivery's problems were, *why* they were having these problems, and *how* the company could fix them. The first part of the presentation included a readout of two reports commissioned by the team: a customer survey and a benchmarking study. The survey results (see Figure 15-3) showed that customers liked the material provided by Comic Delivery and "thought it was a great idea," but customers gave the company poor marks for speed, reliability, and cost of the service. Customers were frustrated with the delay between placing an order and the actual delivery of a joke. Generally, they wanted to have a joke delivered as soon as they placed the order. More important, however, they wanted the joke to arrive when and where it was promised. Often, the person who

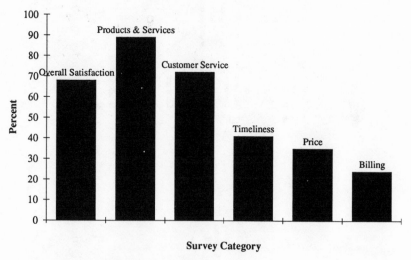

Figure 15-3. Customer satisfaction survey.

placed the order made elaborate arrangements to have his or her target at a certain place at a certain time, only to be disappointed when the delivery did not take place as promised. Last, customers complained that the cost simply was too high, and said that they would use the service much more frequently if it were priced more reasonably.

The benchmarking study showed that other companies were able to do better in some areas and that Comic Delivery's problems were not endemic to the business. For example, smaller start-up companies were able to deliver service faster and for less money, though they did not (yet) have the same quality products as Comic Delivery.

The second part of the presentation focused on why the problems existed. The team used the methods outlined in the "hunt and gather" phase to identify root causes and to understand the dynamics of the current process. Based on interviews with operations personnel and management and on analysis of the current process and systems (see Figure 15-4), the team identified five root causes of the order fulfillment problems:

1.   The order entry was long and entirely manual.

2.   The entry of status information was manual and done in batches.

3.   The process was fragmented; groups were optimizing subprocesses such as order entry and scheduling instead of focusing on the extended order fulfillment process.

4.   The systems used different hardware and software architectures.

5.   The process of recovering money by issuing bills, especially bills for small amounts, was inherently slow and costly.

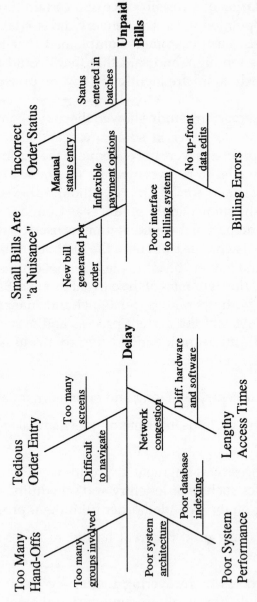

Figure 15-4. Results of root cause analysis.

Using simulation, the team was able to quantify the behavior of the process. For example, they found that the average time from order to delivery was five working days. Yet, the actual work involved in delivering service took only 135 minutes, most of which was the physical delivery of the material. During the rest of the time, the order was sitting in someone's in-basket or was being reworked because the driver went to the wrong address the first time.

Running the model with different order arrival rates, the team found the throughput of the operations to be much lower than expected; that is, the operations only could handle a thousand or so orders per month. If business was good during a particular month, a backlog was created, and orders took longer (on average) to fill. Figure 15-5 illustrates a common mistake in measuring processes. The cycle time—the time required to fill an individual order—seems to vary wildly. As the cycle time varies, managers make adjustments in the process; but nothing is changing in the process. It is only when the number of orders exceeds the capacity of the operations that a backlog is created and the cycle time statistics are adversely affected. The throughput, however, remains fairly constant and does not exceed an upper bound.

The model also provided the team with a detailed understanding of the dynamics of each job function performing parts of the process. More precisely, the model quantified resource utilization: how much time was spent by each job function (sales rep, delivery person, etc.) doing actual work, doing rework, handling non–value-added tasks, and so on. The resource utilization charts are shown in Figure 15-6. Note that the sales rep is spending less time talking with customers than doing non–value-added work, such as faxing, photocopying, or answering management queries about order status.

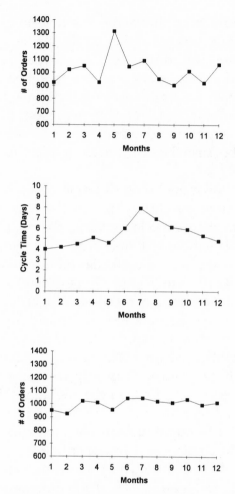

Figure 15-5. Monthly order arrival, cycle time, and throughput.

At this point in the presentation, the reengineering team members had convinced their audience that they understood Comic Delivery's problems and why those problems existed. After gaining credibility, the team presented a number of ideas on reengineering the order fulfillment process.

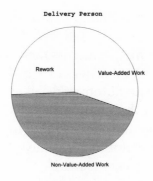

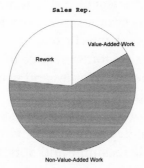

Figure 15-6. Resource utilization charts from the simulation model.

REDESIGN CANDIDATES

To generate ideas, this team focused on the root causes and how to eliminate them. They came up with dozens of ideas, many of which represented radical departures from the existing ways of delivering comedy:

- Eliminate billing by offering discounts to customers who prepay by using a credit card.
- Give couriers devices to let them check status and changes to orders on-line.

- Eliminate some or all in-person deliveries by using automatic dialers to "deliver" the material by telephone or media such as facsimile and mail.
- Eliminate manual order entry by having customers interact with an automated call distribution system, thus entering their own orders.
- Eliminate manual status by having delivery personnel use mobile computers to update order data as the delivery is made.
- Reduce the time it takes to select material that meets a customer's request by consolidating the multiple, heterogeneous systems into a single logical system with a single user interface.
- Provide better customer service by maintaining customer ordering profiles and using these data in promotional activities.
- Consolidate job functions to reduce hand-offs.
- Reduce data entry errors by providing automatic order entry via facsimile and optical character recognition.

It is plausible that each of these ideas could improve Comic Delivery's order fulfillment process. Yet, it would be too expensive to have a trial of each of the ideas. In addition, some ideas may have a greater impact than others, or one idea may even cancel out the benefit derived by implementing another. How can Comic Delivery determine which way to proceed, based on the reengineering team's analysis?

Instead of employing the typical SOTP (seat of the pants) method, Comic Delivery's reengineering team used the methods and tools described in the "innovate and build" phase to iteratively select and then validate elements of its redesign.

THE REDESIGN PROCESS

Throughout the redesign process, the team members should view themselves both as explorers and as scientists. As we emphasized in Chapter 10, it is not possible to create a perfect design on paper and then proceed directly with implementation. Rather, you will use the simulation models and sound experimental design to select design candidates from which you will benefit most. Then you will use evolutionary development methods, enabled by the key technologies described in Chapter 12, to implement a subset of the new design and verify that it produces the expected benefits.

Comic Delivery's redesign process (shown in Figure 15-7) will not apply to all reengineering projects, but it contains the basic elements of a successful redesign process. This iterative process has eight steps, which are described below in further detail:

1. Create a multifunctional team that represents the organizational assumptions included in the new design.

2. Design an experiment to compare the different design possibilities.

3. Create simulation models for each combination of design elements in the experiment.

4. Select a process prototype candidate based on the results of the experiments performed with the simulator.

5. Create an on-line description of the process using the work flow software.

6. Implement the screens and database accessors using application generators.

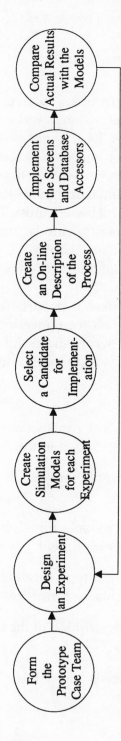

Figure 15-7. The redesign process.

7.   Compare process measurements generated by the work flow software to those measurements predicted by the model.

8.   Iterate (back to step 2).

Creating a Multifunctional Team

As Hammer and Champy point out in *Reengineering the Corporation*, the functional division of labor that Adam Smith originally employed for the sake of efficiency is now a major cause of *in*efficiency. In this step, the Comic Delivery reengineering team co-located the customer representatives and the delivery scheduling personnel into two teams of four people, each with three reps and one scheduler, which were labeled "prototype case teams." The term case team was used because it described a multifunctional group that supported an extended business process. "Prototype" was used to denote the evolutionary nature of the work performed by the teams. In lieu of rigid methods and procedures, this team might be changing its process and using different technologies every few weeks or even days.

From the very beginning, the case team personnel were an integral part of the redesign process. Instead of there being engineers and M.B.A.'s directing the operations staff, the situation was reversed. The operations staff members were treated as experts, and their experience and ideas played a key role in the project. The prototype case teams worked closely with reengineering consultants and systems developers throughout the redesign process. In general, the case teams' cooperation and enthusiasm for evolving the design are critical to the success of realizing the vision of the reengineering effort.

Co-locating the different functions within teams physically placed next to each other forced the people on the teams to confront any problems they had with one another. In the past, if a customer representative did a poor job in creating an order, the schedulers would "reject" the order and send it back. They would not allow the poor work of some other group to affect their performance measurements, even if it meant adding days to the process by sending work back. After the teams were formed, however, the *team* was measured. A poorly written order now meant that the scheduler would lean over a desk and question the customer rep directly, not as punishment but to improve the work.

In our experience, simply bringing people together eliminates much of the finger-pointing between job functions and improves the throughput of the case teams by as much as 40 percent when compared to equivalent staffing in the rest of the operation. (This is discussed further in Chapter 16.)

Designing an Experiment

In this step, the team identified which factors should be included in the experiments, which values should be used, and how many trials should be run. Typical factors include staffing (i.e., changing the number of people in each function), whether or not tasks are automated (i.e., changing the distribution of average time to perform a task), order volume (i.e., distribution of order arrival by type, location, and customer), and so forth. The team selected the six ideas they believed would have the greatest effect on the root causes of order fulfillment problems. Using the techniques described in Chapter 11,[1] the reengineering

---

[1] The team also used the experimental design references listed in the Suggested Reading section.

No. of Sales Reps. per Team	Single User Interface?	Automated Database Queries?	Automatic Call Routing?	Average System Resp. Time	Accept Credit Cards?
3	No	Yes	Yes	5 seconds	Yes

Figure 15-8. Values for one of the sixteen experiments.

team designed sixteen experiments, each including a variation of the six design ideas.

A sample experiment is shown in Figure 15-8. There is no magic involved in selecting factors to include in the experiments. The reengineering team selects them according to its collective experience and knowledge of the business.

The iterative nature of the redesign process allows you to recover from mistakes made when you select the factors. If you choose factors that will have no significant effect, you may select additional factors in the next round of experiments. Similar logic holds for the values you choose for the factors. That is why we have called it "traveling through the process space." The results of your first round of experiments only set you in the general direction of an optimal design. The next round gets you closer to the optimum, and so on.

Creating Simulation Models

The experimental design is only a guide. To answer questions about how the process will behave (e.g., will this combination of factors improve throughput?), you must actually run the experiments. Trying to execute and control such experiments in a production environment is impossible. Just generating enough data points to have a valid experiment would take months, given a five-day process. Instead, the Comic Delivery team used simulation, where a five-day process could be simulated in milliseconds. Using commercially available software (e.g., SES's Workbench or the Exerciser in NCR's ProcessIT), the team created sixteen different models, one for each combination of factors in the experimental design.

To build the models, the team had to make assumptions about the detailed characteristics of each factor. For example, in modeling the prepayment option, the team had to decide how many people were likely to use that option and how long it would take to process a credit card while the customer was on the phone. The team also hired a statistician to help it appropriately apply distribution curves and randomization.

Many people are skeptical of simulation models. They ask, "How can you build a model of a reengineered process if all you have are assumptions?" The answer is that building the models formalizes the assumptions you normally make (but usually internalize). Then, a computer can take those formalized assumptions and execute the process, demonstrating the consequences of the assumptions. Of course, a model is not a replacement for reality, but it is an invaluable aid in making decisions when reality (e.g., trying all possible experiments in a production environment) is not an option.

Selecting a Prototype Candidate

The experiments identify the effects of particular factors in the candidate designs. They quantify the effect of the individual factors as well as the interaction effects *between* factors. Thus, the reengineering team can use the information about the effect of a factor or a set of factors, along with information about the difficulty and the cost of implementing changes to those factors, to select a prototype.

Before the models were completed, the team members' intuition told them that consolidating the different systems onto a single hardware/software platform should be the first prototype candidate. After all, that could reduce the search time by more than 75 percent according to the developers, and it would reduce the number of errors associated with incorrect searches. By examining the models, however, the team realized that it was actually the schedulers who were the bottleneck in the process. When experiments were run at the reduced search time, there was no improvement in the throughput of the process. The team only would have succeeded in getting more orders to the bottleneck, where they would have had to wait until the schedulers could get to them.

After looking at the results of the experiments, and balancing the cost, difficulty, and benefit of each design element, the Comic Delivery team decided to implement two elements of its new design: prepayment by credit card and delivery of material by multiple media. These particular elements were of relatively low cost and risk (as compared to the systems integration design elements), and the model showed that eliminating bills would eliminate 80 percent of the source of billing errors.

Creating an On-line Description of the Process

Now that prototype candidates were selected, the Comic Delivery team was ready to put something in the field. They first installed a small UNIX server running a client-server software architecture. (Many of the vendors providing the application development tools described in Chapter 12 also provide client-server platforms.) Then, they made sure the server was accessible from the existing PC workstations and terminals. By using a server instead of one of the existing mainframes, the reengineering team's developers had more flexibility in controlling the environment, while the server acted as a buffer between the reengineering team's development and that of the other production systems. Team members knew that when they needed more capacity, they could add another CPU or server or even port the application onto another machine supported by the software architecture.

This server allowed the team to deploy software to tie the different job functions together and to "wrap around" the existing systems (e.g., by providing a single user interface but still communicating with the existing systems). The reengineering team used the work flow software described in Chapter 12 to tie the job functions together. All tasks were included in the on-line definition of the process. The Comic Delivery team defined both the new design elements and other parts of the current process. Where there was no system support of a task, the team used "stubs"—tasks that represented work occurring outside of the system. Even with stub tasks, the on-line process was defined with the same branches and parallel subprocesses that were used in the simulation model.

By defining the entire process on the small server, the reengineering team members achieved three goals:

1.   They took their first step, moving from pure analysis into the production environment.

2.   They got feedback from the case teams on the work flow software and the implementation of their design ideas.

3.   They were able to generate their own process metrics.

Using Application Development Tools

As noted in Chapter 10, the prototype is not intended as a "throw-away" development. Throw-away prototypes involve creating a mock-up of the change, demonstrating it to users, and then creating the production software from scratch. In an evolutionary approach, the technologies described in Chapter 12 allow for quick implementation and demonstration of the change *on a stable, proven architecture*. The screens are not just for demonstration but are linked to accessors that interact with an actual production database. The case teams thus are doing production work in the new environment.

In implementing the change, "stub" tasks in the on-line definition of the process are changed to tasks that are linked to services. These services could be screens that update the database, could be software that retrieves information from external systems, could be algorithms for calculating pricing, and so on. Thus, functionality is being iteratively layered into the framework provided by the work-managed process. Such changes are made by updating the on-line definition of the process.

Comic Delivery wrote all of the screens supporting the two prototype candidates using screen generators that also generated code linking the screens to database accessors and other services. In support of the new prepayment option, for example, the team developed a new screen to capture the credit card information. The new screen was tied to a communications script that dialed the appropriate credit service and returned an approval code. The screen was generated independent of the production application for demonstrating the user interface to users. After the communication script was tested, a screen was linked with the script, and the on-line definition of the process was updated to call up the new screen at the right time in the process.

Comparing Actual Process
Measurements to the Model

The simulator generated process measurements using assumptions about the process such as arrival rates and the time required to complete each task. The work flow software also can generate process measurements, but they are based on the actual execution of tasks in the process. Thus, the simulator is used as a predictor and the statistics generated by the work flow software are used to validate the predictions. If the measurements do not agree, the reengineering team must check both the assumptions in the model and whether the change was implemented as prescribed in the model. Based on this analysis, either the model or the implementation of the design element should be adapted.

In Comic Delivery's case, the reengineering team assumed that it would take less than five seconds from the time a user entered data until an approval code was returned. However, the actual process metrics showed that completion of this task in the process was taking an average of thirty seconds, and sometimes as long as four minutes. After talking with the case teams and the developers, it was clear to the team that the cause of the delay was the way the communications script was calling the credit services. After some research, the team decided to install direct lines to the major services. This resulted in average task times of three seconds, with a standard deviation of one and one-half seconds, and the models were updated with the new information.

SUMMARY

Seven months after their presentation, the team had successfully installed six major elements of its new design.

The throughput of the prototype case teams was more than seven times that of other comparably staffed work groups, and, with the new means of delivering material, the cycle time was less than five minutes in many cases.

As the pilot was expanded and other case teams were formed, the reengineering team presented the key elements of the pilot's success:

- Having developers, reengineering experts, a statistician, and two multifunctional teams work closely together provided everyone with a rich learning environment that allowed progress to be made quickly.
- Starting off with a small step allowed the teams and their management to gain confidence and to show results.
- Using simulation and experimentation was critical to making intelligent decisions.
- The use of new technologies such as client-server architectures and code generation tools enabled the team to develop prototypes quickly while remaining in a production environment.
- Iteration allowed the team to validate its decisions one or two at a time instead of trying to make all decisions prior to any implementation.

Even with a successful pilot, many organizational and training issues remain unresolved, and there is still the problem of transitioning the solution to the rest of the current environment. These topics are discussed in the "reorganize, retrain, retool" section.

Phase *IV*

Reorganize,
Retrain,
Retool

In school we all were taught the three R's, which have long been recognized as the foundation for primary education in the United States. If the three R's are not satisfied, the educational system, the student, or both have failed. A similar situation holds true in reengineering. Reorganize, Retrain, Retool are a different set of R's, but their satisfaction is a similar barometer for success. Frequently, this final phase of a reengineering project is labeled "transition" or "integration." These may be correct labels for *what* you are doing, but the three R's represent the *how to* for this phase of a reengineering project. As in the "hunt and gather" phase, which looked at all parts of the problem, the goal in this phase is to integrate the solution—processes, organizations, and systems. The following three chapters explain the final phase of a reengineering project and, indeed, are applicable to the implementation of any large project.

REORGANIZE

Businesses constantly reorganize. For example, groups are moved to different divisions because of functional realignments, or downsizing necessitates that more functions be moved under a single executive. In these contexts, reorganization is done to "shake things up" or in reaction to a new strategy. Such concepts are foreign to the reorganization approach that reengineering demands. Reengineering requires that the organizational model be based on the new business process.

Chapter 16 provides guidelines for organizing around the new processes and information systems. Much of the success of reengineering depends on reinventing how work is done and the role of management in supporting the work environment. The new process needs to be effected with a new organization—different in structure, skills, and culture. A variety of organizational concepts are especially applicable to the organization created by reengineering. For example, the deployment of high performance or self-directed work teams may be used to move to a process-based organization.

A well-developed reorganization plan can be sabotaged, or in some cases carried out unprofessionally, if one key area is not addressed: the management of the inevitable staff reductions, where people are dismissed from the company or hopefully (if this is done properly) reassigned to a different part of the business. There are dozens of wrong ways to handle these situations. This chapter shows how to avoid wrong ways and perform the job at hand professionally.

RETRAIN

How many times have you heard it said that "people are our most valuable resource"? This adage begs the question, "How many companies or organizations have a good grip on what their employees are capable of doing?" The answer is that probably very few do. Thousands of employees work in jobs where their individual skill sets do not match the demands of their job assignment. Reengineering requires that a business get at the root of this problem. An in-depth understanding of the capabilities of the employees is essential to proceeding with the retraining needed to make employees productive in the new environment.

Despite its importance, training often is the most ignored and misused tool in many companies. A common flaw is that people simply are not trained in the things that will enable them to do their job. If they are trained at all, it is for a myriad of programs and technologies that usually do not apply to the activities that they must perform on a daily basis. The result is that much of the money spent on training today is completely wasted. Reengineering projects cannot afford such waste. Chapter 17 should be used to formulate a training plan based on the work that needs to be done. It introduces nontraditional, nonclassroom training programs that have proved successful.

RETOOL

After the process is documented and the training is completed, it is time to introduce the new technology to the rest of the organizations and operationalize all of the new work groups. The wide-scale cutover to a new information system must be planned prudently in order to avoid risk and increase the chances for overall success. The transition to the new work groups is equally important. In many ways, it is as if you are reopening your business after some significant renovations. After all of the experimentation is complete, everyone and everything must be in the right place when the new business is unveiled. Chapter 18 explains what it takes to smoothly expand the reengineering solution beyond the prototype case teams.

16

Reorganize

Businesses reorganize continually. We all have listened intently to hallway rumors about the latest reorganization that is in the works. However, a reorganization usually has little net effect on the daily work activities of most middle and lower management employees. Instead, they regard reorganizations with either skepticism or fascination. A balance exists between sarcastic comments about how "management has no real clue what to do" or that "they just want to shake things up a little bit" and the soap opera aspects of guessing what will become of the current executive. Because this is the "normal" paradigm, the mere suggestion that reorganization is essential to a successful reengineering project causes some apprehension. A little of an "Oh, not again!" reaction is perfectly natural.

Yet reorganization is necessary to satisfy reengineering

objectives. If the business process truly is reengineered, it simply will not work within the old organizational structure. If we remember the fundamental purposes of the old process, it is easy to recognize why it is necessary to reorganize. Old processes were designed around functional organizations (function *first,* process *second*), and supporting information systems were designed in a similar fashion. Systems were designed to support the goals or the objectives of functional organizations. The process always was a secondary consideration. The entire purpose of redesigning the business process is to create a more fluid, unrestricted process that is supported, not constricted, by information systems.

As opposed to typical reorganizations, which involve either the realignment of functional organizations by adding or deleting subfunctions of the organization or the "revolving executive" syndrome, the reengineering focus is designed to create an organizational structure that can best support the business process and have built-in mechanisms for continuous improvement. Designing organizations around revamped processes yields excellent opportunities for us to utilize the types of organizations that are espoused as goals of quality improvement initiatives. For example, high performance teams can be created out of necessity, rather than as part of a long list of goals for the quality effort. Leveraging the process redesign into breakthrough organizational structures parallels the notion of leveraging technology to support the new process.

This chapter serves two purposes: we discuss an approach to reorganizing for the reengineering project, and we offer some insights into problems that you will face. There is no magic formula for creating the model organization. There are, however, some common ideas that are central to "world-class" organizations, and these ideas should be used as guidelines for the reorganization effort.

A REORGANIZATION CYCLE

The reorganization should be a by-product of the concurrent reengineering model. The optimum organizational structure should be found by running trials of new organizations simultaneously with process changes. In some cases, however, such as those where the information technology platform is immature, a more protracted transition phase may be required. In addition, careful consideration must be paid to each of the areas discussed below even when the concurrent model is plausible. This is especially true in deciding what management structure best supports the new process. The concurrent reengineering model has been used most effectively at the "working" or task level. Nevertheless, it is also possible to have trials of changes to management structures when process designs are approached iteratively.

During the "innovate and build" phase, you may decide to use temporary working-level teams to perform the analysis and redesign as recommended in Chapter 9. Cross-functional teams are required to accomplish the concurrent redesign of the process and systems, but, from an organizational viewpoint, the formation of a cross-functional work team is only a starting point. You must evaluate several issues before simply implementing the multifunction team approach throughout the entire process. First, most concurrent process redesign efforts are accomplished by using a microcosm of the total workforce. The challenge is to effectively install the optimal work team structure throughout the organization. Second, it may be that functional tasks that are being carried out in the existing process by one or more groups are no longer value-added tasks in the redesigned process, or the tasks themselves may be absorbed by the cross-functional team because of improvements in the information systems. The major areas that the reorganization design and implementation fall into are:

- Task identification and mapping.
- Choosing the "team" structure.
- Choosing a management structure.
- Implementing the reorganization.

TASK IDENTIFICATION AND MAPPING

The purpose of this activity is to identify all the value-added tasks in the process and map them to functional skill levels and ultimately to workers. Before getting out your favorite PC drawing package to create organizational hierarchies with names filled in, you first must identify all the tasks that will be performed and get some idea of the skill levels required of the people who will perform them. To do this, use the task descriptions that are outputs of the "innovate and build" phase. You will have gotten an idea of workers' collective experience and of the skill levels required by experimenting with the redesigned process and the prototype case team. Two primary questions arise once a target task-to-worker assignment has been completed. The first deals with whether, through training, personnel can meet the skill requirements for the job that is defined (this question is discussed in Chapter 17). The second is the matter of staffing levels. For instance, how many component testing people are needed on the radio assembly line? Or what is the best ratio of component testers to dial installers in a multifunctional team? Process simulation plays an integral role in the second area.

An accurate simulation model of the redesigned process will help drive decisions about staffing. The approach is twofold. The model is used to find the right "mix" for a team. After modeling the redesigned process to get a handle on the throughput and the cycle time, you should use the model to find the optimal team composition in terms of numbers and ratios. The goal is to find the team dynamics that produce the fewest bottlenecks in a process. Of course, factors such as process load are critical inputs to this determination. A simplified way to think about this paradigm is to remember the old "I Love Lucy" television show when Lucy was working in a candy factory making bonbons. Everything was going well for Lucy as long as Ethel was keeping up with and packing the candy Lucy was putting on the conveyor belt as it moved at moderate speed. Lucy was able to dip the nougats in chocolate and get them wrapped. As the scene continued, the conveyor belt sped up. Both Lucy and Ethel were no longer able to keep up with the process. Lucy was shoving bonbons in her pockets and in her mouth. The scene was a hysterical one, but it teaches us a valuable lesson because all processes have the same characteristics as the candy factory: Work will quickly pile up on someone's desk if additional work comes in faster than the person's ability to perform the task. Also, work will pile up if the person performing a predecessor task completes her work at a faster rate than that at which the next person can complete his work. (The dynamics get much more complicated when a number of people are required to complete several tasks in the process at different times.) The use of a simulator is essential to predict process dynamics under these conditions.

The second part of the analysis can be completed with a calculator. Once the exact team makeup has been decided upon and a sensitivity analysis of throughput and cycle time completed, the number of teams (and therefore the total staff count) is based on the expected input rate to the process. For example, if the input to the process is customer orders, how many are expected? The simulation results will tell you at what order volumes the process will work best and at what point its capacity will be reached. Thus, there is no need to guess at the proper staffing levels, or to do back-of-the-envelope calculations based on limited information. We have found this idea very appealing to management. Managers are frightened by reorganizations because of extra costs incurred after the reorganization is completed, as these costs frequently take the form of overtime due to improper staffing projections. A simulation tool that helps predict what will happen is extremely powerful and must be used.

Elimination of Unneeded Organizational Functions

A problem in many processes is the use of ad hoc organizational structures, which usually are formed as checking or control groups. When processes do not work very well, management frequently responds to problems by putting a new group in place to check the output of one work group as it proceeds to another work group. These control groups sometimes are referred to as inspection groups. Inspection is a well-known facet of manufacturing processes, but in service processes inspection has been done in a rather casual way. Because many service processes tend to spill across several organizations, inspection groups tend to pop up in many places without any real strategy, and ultimately these groups become "part of the process." The process redesign will eliminate the need for these groups. These functions simply do not add value and must be targeted for elimination.

Two other groups of people fall victim to good redesign. The first category is workers who have become unnecessary because of the redesign. For instance, the introduction of information technology directly to the sales force where salespeople can write customer orders and have them immediately downloaded into the system eliminates the need for a group that once existed solely to take customer orders. This is not just brute-force automation, but rather the correct use of information technology to enable the redesigned process to become a reality. The second category includes people whose functions have been absorbed within the multifunctional team that will support the new process. Determining what functions can move within the team is discussed in the next chapter. The goal is to collapse as many tasks or jobs into as few people as possible while still maintaining high throughput.

Once these superfluous groups have been identified, a strategy must be developed to communicate why and how the changes are going to occur. Reengineering projects create a lot of tension for everyone. Everyone knows that change will occur, but the uncertainty of what the change will bring can create fear. Many people welcome change, but more resist it. In many ways, the identification of groups that no longer are needed marks the first time when it is obvious that things really are going to change. Here the change management program (supported by information provided by the public relations person on the reengineering team) kicks in. Employees within these organizations must understand why the functions they currently are performing no longer will be performed by them, if they are performed at all. These people also may be at risk, depending on how important the skills required to complete their tasks are to the business. From a human perspective, this is the most difficult part of reengineering. For everything that can be gained for the corporation in the long run to keep it competitive, there is still the short-term problem of dealing with potential layoffs or at least the movement of employees to different jobs. This problem deserves special attention, and a later section discusses it at some length. It is important at this point to understand that this is inevitable. The situation must be dealt with in an uncompromising way.

CHOOSING THE "TEAM" STRUCTURE

There are multiple paradigms for team structures. The latest literature in this area discusses the concept of high performance teams. The underlying premise of high performance teams is the idea of empowering work groups so that they can act as autonomous, self-managed units. The productivity of these teams is much higher than that of typical functional units with authoritarian management. In this section we survey some useful ideas for developing such teams.

Many companies have implemented multifunctional teams. Concurrent engineering projects have effectively used such work team structures to achieve quantum breakthroughs in areas such as time-to-market. For example, engineers, salespeople, and manufacturing employees work as a team to conceptualize, design, test, and deploy a product. End-to-end ownership of the product development cycle creates pride, accountability, and less chance for rework than there is in a waterfall model. (Ownership is discussed further below, in the section on "Choosing a Management Structure.") Such teams, however, tend to be ad hoc structures created for the purpose of developing a product. Team members usually report to their current management structure or perhaps have some type of matrix reporting structure, so that these structures are in a tenuous position. As long as management supports the concept, they will remain intact. However, pressure from the functional organizations may cause their resources to dry up. Therefore, these "nonpermanent" team structures should be replaced by permanent organizational structures.

As discussed in Chapter 9, during the redesign phase a group of workers from various functional organizations are assembled to make the analysis and prototyping work easier. This work team approach should be the basis for the permanent team structure. You will quickly discover that tremendous benefits are realized only if the teams become empowered. In spite of all the work done by Deming and others on driving fear out of the workplace and giving workers the chance to take pride in their work, management is still struggling with empowering employees. Applying a team approach during the redesign phase provides an excellent opportunity for a demonstration that such a structure can work. This makes the transition to empowering all teams a little easier. So, we need to determine how high performance teams can be established to function effectively.

The greatest barrier to establishing high performance teams is the lack of committed management. As mentioned above, the use of this type of team during the redesign phase on a trial basis demonstrates its true effectiveness. Fortunately, if top managers really commit to reengineering, they probably have the intestinal fortitude to make this happen also. The hallmark of the team is empowerment, which is not meant to have a placebo effect on the workers—it just makes good business sense. The principle is to keep management out of the value-chain work flow as much as possible. One premise in reengineering processes is to design a process with only value-added steps. Having process steps that require management to inspect, check status, or control is one sure means of screwing things up once the process is deployed. These teams can function quite effectively where an up-front partnership is formed between management and work teams to delineate where the span of control begins and ends.

One way of thinking about the idea is in terms of

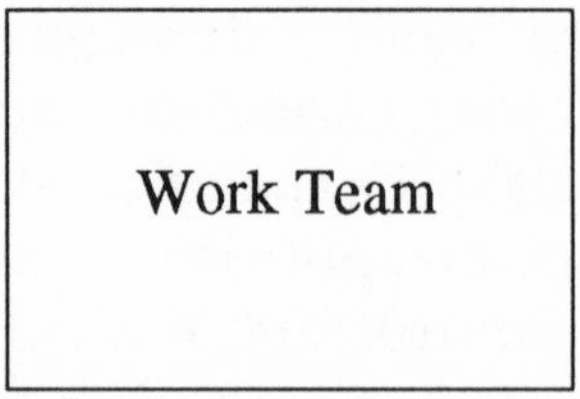

Work Team

Figure 16-1. Organizational view of boundaries.

boundaries. If you take a piece of paper and write the name of the work team in the middle and then draw a box around it, a boundary has been created (see Figure 16-1). The exercise that follows is intended to delineate responsibilities that belong in the box (workers) versus those responsibilities that are outside the box (management). There is no limit to what can be done here. We know of teams that were made responsible for their own vacation schedules, yearly reviews, and personnel decisions (at least firing). Defining which responsibilities are acceptable for the team and which for management is purely situational; factors such as the culture, corporate administrative policies, and comfort levels are all considerations. A caution is to make sure that the team does not become so overburdened in administration that all the process benefits are swallowed. The effects of this overhead can be measured by using the process simulator. Figure 16-2 is an example of the output.

A question that frequently arises concerns whether the team members must be co-located. It is fallacious to think it will suffice to put people in an organization (whether functional or not) and deem the collection of names on an organizational chart a "team." Announcing that there is a "transition team," or a "reengineering team" for that matter, is absolutely the wrong way to do it. (How many football teams do you know that do not play on the same field together?) Without question, the best organizational

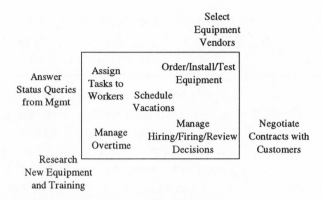

Figure 16-2. Responsibility view of boundaries.

structure is one where all people working to create the same output (e.g., customer orders, insurance claims) are also in the same location. This is difficult in extended processes that require workers to be in the field (e.g., the sales force in an order fulfillment process impacts an assembly factory). However, even in these cases, the team members must spend enough time together to build the rapport that humans need to interact in an effective manner.

Building effective teams is most challenging, given the work environment of the nineties. The introduction of technological developments such as electronic mail and telephone answering systems has diminished teamwork tremendously. Face-to-face meetings have been replaced by impersonal electronic mail systems. A technology that was created to keep people in touch has been so abused that many managers spend as much as half their day just reading and responding to electronic mail. At the non-management level, functional work groups usually are geographically dispersed and forced to communicate via telephones. Instead of talking to humans on the other end of the line, they usually get a telephone answering system. These technologies have helped create a "people avoidance" factor that works against the team concept. As companies have constantly decreased travel budgets to reduce costs, the opportunities of personal interactions have declined. The point here is that an idea such as co-location is very simple, but many companies have moved far away from it. The challenge is to capture the spirit of teamwork again. The best way, as we have said, is to put people under one roof. Another way is to plan and organize more face-to-face meetings among team members. These meetings must be working sessions instead of periodic status or planning sessions. Although most companies do not possess all the technologies, one of the workplaces of the future will be the "virtual organization."

The idea of a "virtual organization" is hardly a new one. People mistakenly believe electronic mail or groupware to be the beginnings of linking employees together via electronic media as they work. Our concept is a bit different from this and acts as an alternative to co-location where co-location is not plausible. In Chapter 12, we said that an effective way to use information technology is to put the business process "on-line" within the information system. The software is responsible for the management of tasks. As one task is completed, the next task becomes activated for a person to work on it. Workers are notified that they have tasks to do by electronic work lists. A "virtual organization" extends this concept. Workers in different parts of the country can be linked via the system to perform tasks within the same process in the same manner that they would be performed if they were co-located. We believe that productivity still will suffer if the workers are not co-located, but the implementation of multimedia technologies such as interactive video on the same workstation will make these situations more workable. In most cases, multimedia technologies are not embedded in the infrastructure of corporations. When redesign is contemplated, it may be less expensive to invest in such technologies than it would be to move workers from state to state.

Some companies have had success in extending the team paradigm one step further. Instead of using a collection of people with multiple skill sets, these companies have implemented case workers—a case worker being a single person who owns a process output from end to end. We believe that case workers are a natural progression from a case team environment. As information technology gets more powerful, individual workers will be able to perform more tasks without creating process bottlenecks. Of course, all the principles described above apply equally well to an environment where a move to case workers is plausible.

CHOOSING A MANAGEMENT STRUCTURE

In *The Wisdom of Teams*, Katzenbach and Smith define a team as "a small number of people with complementary skills who are committed to a common purpose, set of performance goals, and approach for which they hold themselves mutually accountable." We all may have had grand ideas on how we could accomplish work better or may have been part of a reorganization that seemed to be an improvement over the old one. Yet, what ruins many good intentions is actual implementation. The first thing that typically does not change is how the business or the work is managed. The authoritarian people remain the way they are, as do the care-free ones.

To make a reengineering implementation work to its fullest potential, there must be a wide-scale change in how the process is managed. Obviously, it would be best if management style changes were made at all levels in the hierarchy, but we will focus on the first levels of management to be affected by reengineering solutions.

To make any type of high performance team structure work, managers first must unlearn much of their behavior and change their focus entirely. The control paradigm must be changed to the facilitation paradigm. Management techniques must be replaced by forward-looking thinking and leadership. High performance teams must be given the freedom to succeed. Management must provide an environment where they can flourish. In hierarchical terms, there are two models for the level of management whose responsibility it is to support the team. The first is that of the superspecialist. Most people who "grow up" in companies generally gain expertise in one or two functional disciplines (e.g., billing, marketing). If the organization desires that management supporting the team as a superspecialist, the manager will have to be a "specialist" in each of the functional disciplines found in the team. Without this level of broad experience, specialists in each of these areas must be available within the process. The second model is the facilitator or coach. Here, the manager helps the team by making meetings more focused, providing encouragement, and protecting the team from outside interference that is beyond their scope. It will be easier to see which of these two basic models is appropriate in a given situation after a skills inventory is completed, as discussed in Chapter 17.

At the next level of management, the manager must function very much like a leader. To briefly explain this, we use an example from Stephen Covey's *Seven Habits of Highly Successful People.* Covey describes an expedition through a jungle, where the manager ensures that the workers are cutting down the branches ahead of them in an expeditious fashion and that their machetes are kept sharp. The group's leader, on the other hand, decides whether the expedition crew is even in the right jungle. The point is that a clear direction must be set.

Team managers must be looking forward and planning work in a six- to twelve-month time frame. They must understand the process and the process metrics so that resources are available to handle the workload in the months ahead. In addition, they will be liaisons between their process and other processes in the company. No business processes, no matter how broad, work in isolation. Therefore, they must spend time understanding the over- all strategy and how other processes may affect their own—getting a big-picture view. For example, the order fulfillment process may need to produce output for the product development process and vice versa. Understand- ing how business processes affect one another is a man- agement responsibility that cannot be ignored, or process boundaries will become blurred, and extensive non–value- added information exchange will take place. At the same time, this level of management must not totally ignore the work teams. The manager must make periodic visits to the work environment to remain steeped in reality.

A key reorganization strategy that Hammer professes is that the process must be owned by the process owner. Hammer believes that the process owner is a person who is designated before the redesign and carries it on after its implementation. The process owner is generally a senior executive responsible for the entire process. (The reengi- neering leader may become a process owner.) Ownership in this case means accountability—the person's neck is on the line. We believe that the process owner idea is a good one because it gives a single person the responsibility for what many functional heads used to own. There are some cautions though. The process owner should not have a large organization that is responsible for owning the process. The reaction of many managers is to create an organization around a responsibility, but this must be avoided. Second, the manager must truly own a process. Frequently, we have seen processes confused with prod- ucts or in some cases functions. A nasty problem will occur if there is bottom-line accountability for a process and at the same time a product or a function. These two paradigms do not mix. For example, if Bill is the owner for product X and Jim is the owner for process Y that happens to include the development of product X, then who has overall accountability? "Both" is not a good answer. We argue that if a process focus is chosen, the power of the process owner should transcend that of any product owner who falls within the process.

A problem plaguing many organizations is that of quality control. Most companies have responded to it by forming quality control organizations responsible for the selection of metrics and their measurement. In the reorganized business, the measurement of the process should be moved as close to the action as possible. Therefore, the multifunctional team should have input into the selection of the metrics. If the information system is developed with measurements in mind, they usually can be incorporated and captured automatically. Nevertheless, the responsibility for the quality of the work is that of the team doing the work. To support this day-to-day effort of keeping metrics and process improvement activities, there must be statistical support (e.g., people with master's degrees and relevant experience in the application of statistics) within the organization. In an operations environment, they should be operations research people. In a systems organization, they usually are responsible for performance modeling. The evolution and the continued use of simulation and experimentation cannot be lost in the new organization. Instead, these tools and the people who can use them should remain. This group of people must support the multifunction teams performing work. For example, there may be problems whose solutions require additional statistical background that the team does not possess, or it may be necessary to coordinate process changes across a number of teams. These people also must support any forward-looking work for the organization. Staffing requirements and new business opportunities do not end when reengineering is over.

The factors we have been describing are important in redesigning the organization, but this is a difficult job that requires a great deal of strength on the part of the leader and/or the champion of the effort. Unfortunately, there are problems that never can be predicted during the redesign, but which occur during the effort to mobilize the entire organization for change. The next section deals with some of these issues.

IMPLEMENTING THE REORGANIZATION

The day comes when it is time to roll out the new organization. Always remember that there is no such thing as a smooth transition. When you are dealing with large numbers of people and hundreds of interrelated events with an abundance of permutations, there are bound to be surprises. Depending on the project, there may be an opportunity to use a phased approach. At other times, you may only be able to flash cut. (Some helpful pointers in that area are found in Chapter 18.) For now, let us assume that you have a pretty good plan. There are times when things should be written down—this is one of them. Let us also assume that the plan has been adequately communicated. This is no easy task, but that is why you hired a PR person for the effort and involved some of the "insiders."

Reengineering projects should not begin with the aim of reducing staff by huge amounts. The goal is to design processes that are innovative, are efficient, and help the company make money. Necessarily, a redesigned process will be free of bureaucracy and busy-work functions that exist in a self-fulfilling way. The outcome is that fewer people will be needed to run the new business process. In

March 1993, the *Wall Street Journal* reported on its front page that reengineering would eliminate up to 25 million positions throughout American companies in the following decade. Therefore, it is a mistake to ignore or hide potential staff reductions within the candidate process for reengineering. The threat of downsizing runs directly against the grain of two of Dr. Deming's fourteen points: drive out fear, and create jobs and more jobs. This is the most obvious way in which quality improvement and reengineering diverge.

In fact, the United States is not Japan. America is not starting with a clean slate. Becoming competitive by incrementally improving an outdated process creates more chance of bankruptcy than profits. By now you definitely understand that reengineering is not a random cost-cutting effort designed only to help the company's stock look more attractive in the short term. Anyone can see that downsizing probably will occur within a reengineered process. However, many companies fail in their handling of this painful part of the project. We contend that this can be handled properly, and we believe that corporations have an obligation to do it right. A variety of measures can be taken in approaching downsizing.

Downsizing Strategies

The best possible scenario for all involved in the process is to be able to keep the same amount of staff. This is accomplished most often in situations where a large increase in demand on the process is expected. Not all companies are on the verge of going under when they undertake reengineering. In fact, many have fine products that customers want. In these cases, companies are struggling over how to get the products to the customers or about how to enhance them. The potential for new work then allows the

staff to remain. Of course, there still must be reorganization to meet the redesign.

A derivative of this situation is that the process may
anticipate tremendous demands in the near future. In this
case, employees should be retrained for future use within
the process. This approach works particularly well because
it makes people very cooperative during the redesign
phase, and provides a strong foundation of skilled workers as the process demands grow. In this scenario, people
must not be just put into the day-to-day process even
though they are not needed. Two problems develop. First,
when there are too many people for the job at hand, the
process is not working at optimum capacity. People get
bored if they do not have enough work. Second, extra
people usually find ways to occupy their time, and this
time may be spent in restoring layers of busy work to
the process—"the work will expand to fill the day and
the staff."

At a minimum, all workers should be retrained so that
they can be utilized in other processes within the corporation or gain skills that are desirable for other growing
corporations. Without sounding too idealistic, we believe
this is a social responsibility of the corporation and is a
small price to pay in the overall scheme.[1] Decisive measures
which show people that the company is taking an interest
in their well-being are better than no effort at all. After all,
these people may continue to work for the company or
may be customers who can influence other consumers. Do
not think that walking around with a hatchet is the way to
do things. An often heard excuse is that the information

---

[1] In *Deming Management at Work,* Mary Walton describes how Bridgestone retrained
employees during a lengthy period while factories were being upgraded. The
return on that investment in the form of improved loyalty and productivity helped
Bridgestone overcome the problems faced by its failing predecessor (Firestone).

about the downsizing has come too late into the process for management to take any useful action. This is flat-out incorrect. With the use of simulation and a transition that begins early in the project, there is no excuse for not knowing three to six months in advance that such changes are going to occur. The reengineering leader and the PR person should be out front communicating about the possibilities of downsizing. All questions must be answered. Early on, the answer may be that "we do not know." However, even that answer is superior to avoiding the topic altogether.

There are tremendous psychological impacts of reengineering, too many to discuss here. In fact, there will probably be books about that subject within the next few years. In reading the last few pages, you should have experienced some discomfort. Just the thought of people losing their jobs should hit a nerve. At a practical level, you should be wondering how people will react to such change—their fears, anxiety, resistance, and so forth. Actually, they probably will react as you would expect them to and in ways that you would not expect. Hiring psychologists to assist in this area is a prudent idea, but what we have seen work best is participation. Once they understand that things are going to change, many people will admit that there are inadequacies in the current process. The more that people get to vent their feelings and voice their potential solutions, the easier it will be for them to change. For now, keep in mind that communication must be done early and often. You must encourage participation without compromising the redesign principles.

Relocation

The idea that people will be moved to support the new process should not come as an afterthought. Workers may have to move on the same floor, to different buildings, or even to different states. Work will be needed to create floor plans and to move telecommunications equipment. All plant managers know this. The only reason for mentioning relocation in this book is that it often is overlooked—there will just be the vague realization that people will have to be relocated to make the redesign work. Unfortunately, there will be no plan and most important no budget for relocation. Do not fall into this trap; be sure that money is put aside for this activity.

KEYS TO SUCCESS

- *Do not ignore the management structure.* This was hinted at earlier. Reengineering projects should not become staff exercises where the management figures out a better way for workers to perform labor only to keep their positions sacred. Nothing should be sacred, especially the way that a process is managed.
- *Reorganization should not be conditioned on early successes.* One of the biggest mistakes made is for management to say, "Once you prove that the redesign process does what you say it will do, we will make the changes." The reorganization is part and parcel of the project. It must happen. If you wait until things are working great, they never will. Bad structures on top of good process designs make for poor processes. There should be contingency plans to keep the organization as is but the contingency should never become the first plan.
- *There can never be enough communication.* For reengineering types this warning can be especially troublesome. Just because people were informed once or twice does not mean that the message has

gotten across. The communications must be extensive in depth (why are we doing this?) and amount (tell them a lot). This point cannot be overemphasized.

- *Just because teams are formed, they should not be sacred.* If a team is formed and the personalities are not working out, change the team. A typical reaction is that people need to work out their differences. That may be true in a marriage, but it is not necessarily true in organizations. One way to offset poor team formation practices is through the use of psychological profiles such as Myers-Briggs. The Myers-Briggs reports also can be used to help people resolve minor differences.
- *Do not ignore new organizational paradigms.* The area of organizational development as part of reengineering projects is a fertile new field. We expect a variety of people's experiences to be published in the next few years. We recommend that you use these sources. The emerging area of networks in organizational understanding and development is one such example.
- *The organization must be in a position to improve continuously.* The use of statistical quality control techniques at all levels of the organization will help to ensure that the process continues to improve. We have seen organizations use simulation during the redesign phase but completely throw away the tool once the reengineering project was completed. The tools chosen to perform reengineering are almost always reusable, especially in this area.

17

Retrain

In *The Fifth Discipline,* Peter Senge claims that it is no accident that organizations learn poorly. Senge points to several reasons, but all focus on a type of organizational socialization. A basic premise is that organizational actors become prisoners of the system that they work within. Thus, when situations occur that require a member of an organization to think through a problem, the person generally tries to solve the problem according to learned behaviors developed through his or her experiences within the organization. As time goes on, it becomes difficult to teach old dogs new tricks. People become stagnant. They associate themselves only with their positions, and are unable to understand how their jobs relate to or affect others. Their "mental models" must be changed in order for a reengineering project to be successful. Many opportunities exist to change such mental models throughout a reengineering project, but no time offers a better opportunity than the time used to retrain the organization.

THE TRAINING PROBLEM

Most managers realize that training is essential for an organization to be productive. New hires must learn the business—the policies of the corporation and the subject matter of the work. As new technologies emerge, the existing workforce needs to learn about them before they are implemented within the process. In spite of such common-sense awareness, workers typically are either untrained or poorly trained. If everyone knows that training is important, what is wrong? There are several obvious problems and some that are harder to discern.

A fundamental problem is the ratio of education money spent on executives compared to that spent on "line" workers. During the 1992 presidential campaign, President Clinton stated that approximately 80 percent of corporate education funds are spent on the top levels of management. Education for executives usually takes the form of new management practices taught by very high-priced management consultants. America's competitive position and the skills of most workers are living proof that the small percentage left over for the working folk is simply inadequate. This deficiency is further compounded during recessionary times. Management focuses on running the business more efficiently, which to many managers means reducing costs. Unfortunately, training often is treated like a luxury expense.

In addition to a lack of funds, there generally are problems with the training itself. A common problem with training is that the information learned during a course is not used on the job right away. If newly learned material is not used, there is no reinforcing process to make the information useful. Further, much of the training offered is similar to survey or introductory courses on topics. As little money is being spent in this area, the in-depth training just is not available. Instead, organizations settle for an introduction to many different subjects. As technology rapidly advances and computers automate simple tasks, all that will remain for workers will be the more difficult tasks. These higher-skilled tasks are basic to many reengineering solutions; so, for reengineering to succeed, maintenance of the status quo will not suffice.

A REENGINEERING TRAINING CURRICULUM

Most of the material discussed below is general information that will be applicable to many training needs of organizations. The strategy is to use reengineering as leverage to make changes that would be desirable in any case. The idea that there will be new processes and new information systems is usually impetus enough for even the more stubborn managers to admit that both the subject matter knowledge and the process level knowledge currently in place are inadequate.

An immediate goal is to ensure that the process and the organizational plan that were developed can be implemented. The basic formula for doing this is to understand the current capability of the workforce and to see how this capability stacks up with the needs of the new process and organizational strategy; to assess the gaps; to develop a training curriculum; then to retrain. This plan along with other considerations is laid out in the remainder of the chapter.

FIND OUT WHERE THE ORGANIZATION IS—SKILLS ASSESSMENT

Very often, organizations do not know what their capabilities are. We challenge you to ask the head of your department what his or her organizational strategy is in terms of the skills of the employees and how they match the work. Further, you may even want to ask how new projects are assigned or how people are matched with jobs. The answer probably will be a blank stare. The reason for a blank stare or a rambling nonsensical answer is that few managers have a real clue as to what their organizational capabilities are. Most personnel decisions are made by the immediate supervisor of the person being hired, and any approvals are mere formalities. The process actually ends with the recommendation of the supervisor. In sum, very few organizations have an overall strategy for personnel assignment, selection, and development. Without such a strategy, there can be no clear understanding of where an organization's strengths and weaknesses lie.

The absurdity of this situation is apparent when you consider a football team. Can you really imagine the coach and/or the general manager not having a clear understanding of what each and every player on the team is capable of? Businesses, on the other hand, make decisions all the time without making such an evaluation. Think of the backfield coach just grabbing a player who may be able to play running back but has skills much better suited for being a linebacker. Even more important is the way that professional teams choose players in college drafts. They have to understand the game, where it is headed, what players are available, and which ones will meet the team needs. Successful teams do this well, but poor teams do not. This is very comparable to the situation that businesses face. Corporations need to understand the business they are in, where the market is likely to be in the future, what types of employees are needed to meet the challenges, and which ones they should hire.

Organizations in need of reengineering typically lack strategy in this area, so it is the responsibility of the reengineering team to get a handle on the current organizational capabilities. After all, you cannot retrain for a new process without knowing if it can be implemented or how much training will be required. Such understanding is acquired by performing an in-depth skills assessment (sometimes referred to as a skills inventory) of the current organization. The product of this assessment is a baseline for the organizational capabilities.

Many corporations perform some type of skills assessment, usually for the benefit of the employee in anticipation of career planning. The skills assessment that we recommend is a bit different from most. First, the competencies evaluation must be done with a clear focus on the new process. Second, the evaluation is not done just by the employee. Although the skills assessment is described here in the final phase of the project, it could be performed at any project stage.

Finding the Competencies Required

The skills assessment is designed to explicitly enable the implementation of the reengineered process. Before beginning it, the evaluator must have a clear vision of the desired functional (task) breakdown of the new organization. However, there always are some areas that are peculiar to a particular business, which probably will not change regardless of process changes. Therefore, if time and dollars permit, you may consider trying a preliminary round of the assessment techniques in these "common" areas. For instance, if the company is a telecommunications equipment supplier, it may be reasonable to find out what expertise exists in the relevant technologies regardless of the process. For the reengineered process, the question is what jobs do you expect people to perform at both the nonmanagement and the management levels? Once this is decided, an identification of the desired skills is required.

The point is not to find out if a person has expertise in "Do you know how to fill out the so-and-so purchase form?" That is too low-level, and that exact task may not exist in the new process. So, the challenge is to abstract the skills to a higher-level grouping that explains the general capability. In the preceding case, it may be more reasonable to ask "Do you have knowledge of Company X's widget product and how to procure it from them?" This question gets at the essence of the desired skills—knowledge of the products and the order receipt practices of that company. This type of skill is subject matter skill. Other types of skills are "softer." If you are thinking about converting the majority of your managers to facilitators, you may want to find out how good they are in human-to-human interactions.

Administering the Inventory

After figuring out the right questions to ask, there are a number of steps required to administer the inventory.

1. Physical Packaging
A comprehensive survey is divided into three categories: "soft" or teaming skills, operational or "how do I do my job" skills, and technical or "what I really know and understand" skills.

Soft Skills   For the "soft" skills it is imperative to define the characteristic that is being evaluated. One approach is to list the competency and then ask a series of questions.

Coaching Others
1. Communicates objectives in such a way that inspires a strong sense of ownership with all team members—cohesiveness.
2. Recognizes individual skills and appropriately focuses on team and individual objectives.
3. Gives ongoing feedback that is constructive and objective yet promotes self-esteem of group and individual members.
4. Listens to the ideas and concerns of group and its individual members.

Figure 17-1. Example of a "soft" inventory.

(See Figure 17-1.) At the end of the evaluation the results can then be tallied and scored. This scoring is typically done much like that of questionnaires and in magazines (e.g., 20–45 means very good).

Operational and Technical Skills   For operational and technical skills, we strongly recommend a six-point scale that is tiered with questions or examples that move upward in complexity. For example, see Figure 17-2.

For an example related to technical skills, think of the degrees of increasing complexity in which one would master mathematics as a scale:

Number Systems ➤ Sets ➤ Algebra ➤ Calculus.

The above approach normally works better than a typical five-point rating system for two reasons. First, if it is done properly, competencies will be broken out into distinct skills that require a solid foundation (like learning about number systems before mastering calculus). Second, the use of a six-point scale as opposed to a five-point scale avoids the temptation of rating someone satisfactory (score of 3). As there is no middle ground, it will be clear whether a person needs to improve or is a little better than most.

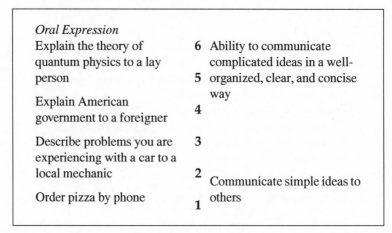

Figure 17-2. Example of operational skills.

We recommend that the inventory be made to look like it was professionally prepared. A black-and-white photocopy of a typed sheet of paper exudes mediocrity. Without creating too awkward a format, pay some attention to the inventory's packaging.

2. Communicating the Purpose

One sure way to sabotage this effort is to fail to tell people that you are planning to do it and why. The PR person and the reengineering leader need to make everyone aware of the skills inventory and why the effort is underway. Honesty is the best policy here. People need to understand that noncooperation will not be acceptable, and that it is in their own interest to supply the best information possible.

To achieve this, the message should be that it is the company's goal to be a world-class leader and have the best-trained people in the industry. Make it clear that reengineering is an important step in that direction. The inventory is designed to provide a realistic evaluation of current skills in order to focus on training for the reengineering effort in the short term and to assist employees in career planning in the long term. Given that the technology life-cycle is accelerating, competition is intensifying, and standards are rising, it is important for all employees to understand where they are. This information will be critical in helping employees to make an objective determination of what they need to do to remain marketable.

3. Finding the Administrators

Sending the surveys out in interoffice mail is totally unacceptable. If people get a letter from a favorite executive with the inventory sheet stapled to it, they will know that this effort is not very serious and probably will not comply with it willingly. The best way to get the inventory done is to have a team of people designated to administer it in the various work groups. This approach serves three purposes. First, it reinforces whatever message the PR person has already offered. Second, it lets people know that the inventory is important and will be done professionally. Third, it ensures that there are people in place to enable feedback to be provided. Most often the skills required to perform the assessments will be found in a human resources department. Either use these personnel or make sure that the people chosen are trained properly. In either case, we recommend that the administrators not be a part of the organization being evaluated.

4. Getting Personal Evaluations

All persons are responsible for using the above surveys to evaluate themselves.

5. Getting Evaluations from Peers

Besides a personal evaluation, each person is responsible for evaluating at least one peer. In turn, each person then is evaluated by at least two peers. The purpose is to get an objective view of how others within the work groups evaluate a person's proficiency in the areas evaluated. As the outcome for most organizations is to construct teams to perform the work, the perceptions of potential members of a team can be as important as the truth. Alternatively, internal or external customers can complete evaluations in lieu of peers.

6. Getting Evaluations from Supervisors

This is another objectivity check. People's supervisors generally look at things differently from the ways those people or their peers do. Also, because a supervisor will be performing an evaluation of each direct report, there must necessarily be consistencies in evaluations across a particular work group for the same competency area. The purpose of using outside, trained administrators is to ensure that there is consistency in the ratings from group to group, an important check and balance precaution.

7. Compilation

Getting all the results and making sense of them can be a painful exercise. For each person, there will be at least four evaluations. Besides brute-force categorization, a careful balance is necessary between reliance on hard scores and "gaining insight" into comments on the evaluations. Comments are the "soft" stuff, and the decisions to retrain in a particular area cannot be made by numbers alone. The more insight that can be gained by examining the results, the better the decisions will be.

8. Feedback

It is the responsibility of the company to share feedback with its employees. A useful way to do this is to have the administrators set up work-group, individual, and supervisory-group sessions to go over results. The power of constructive feedback should not be underestimated. Too often, there is little feedback in the work environment, (e.g., as in end-of-the-year performance appraisals). This evaluation, then, is an opportunity to share information among people at all levels of the organization.

9. Use

After the information is compiled in a reasonable fashion, it must be utilized to figure out if the process goals can be met by retraining the organization. The next step is to perform a gap analysis.

GAP ANALYSIS

Gap analysis is critically important. Given the capabilities of the workforce, we must know whether the organization can be retrained to meet the goals of the new process. The gap evaluation answers two questions: Is the plan feasible? What are the areas to focus on to make it happen? The plan will be feasible if the gap between where the organization is and where the redesign wants to take it is not too significant. Important factors for making this determination are the "base" skill levels of the employees (e.g., are you trying to make computer programmers become airline pilots?) and the amount of time needed for retraining. If you have the appropriate base skills, you can pretty much start thinking about the areas of training that the curriculum must focus on. If not, there are a few options:

- You can always go back and tweak the redesign or the organization plan a bit. The gap analysis is being performed with the output from the simulations as well as the blueprint for the new process. Modifying the simulation model to see where some tweaking can be done was one of the original purposes of using simulation. In some rare cases, you may find out that it is impossible to do what you want. Take a telecommunications example, for instance, where the redesign is totally dependent on a technician taking customer orders as well as performing the task of testing the electrical connectivity. You may find out that 95 percent of the technicians do not currently speak to customers and are terrified to do so. This is one of those situations where it may be appropriate to look at redesign alternatives.

- There may be an opportunity to move people in from different organizations within the company. Done far enough in advance, the gap analysis assists in this type of planning.

After the gap analysis has been completed, a training time-line and curriculum must be developed. The next section identifies the common areas that need to be considered.

CURRICULUM AREAS

In designing a curriculum, attention must be paid to a variety of areas to ensure that the workforce will be adequately educated to perform the process. Of course, there are planning and scheduling activities that need to take place before the implementation. The remainder of this section does not discuss how to develop a plan or what needs to be covered at a course level. Instead, it focuses on the areas of education that must be considered in making the plan.

The cornerstone of organizational development is to build the individual foundation before attempting to develop teams. As the pyramid in Figure 17-3 suggests,

Figure 17-3. The Skills Pyramid.

businesses that attempt to solve personal mastery or team building ahead of subject matter knowledge ultimately will fail. Corporate success is driven by a workforce that has deep-rooted skill sets at all levels of the corporate hierarchy. The remainder of this section describes the basic premise of each level of the pyramid.

Subject Matter Expertise

To paraphrase Dr. Deming, "working hard is not enough, you must know *what* to do." So often, workers are deprived of an understanding of what the business is all about. The focus is constantly on the task, and rarely on the big picture—how the task relates to others and the whole purpose of the business. It is always bewildering to see workers (not just hourly workers) plodding along without a clue as to how their function fits into the business strategy, why it is important, or how it improves profitability for the corporation. The "I am my job" syndrome is most prevalent. *Out of the Crisis* gives many examples of Deming asking employees what their job is. The responses are typically, "I manage the widget production." Deming always responds, "I asked what do you *do?*" Unfortunately, the bottom line is that a large portion of the workforce does not understand the business that the company is in.

To attack this deficiency, focus on getting all employees educated on the business and, most important, how their job relates to the customer (e.g., how the turbine engine works in the plane or how processing a service order fits in with the provisioning of a telecommunications network). A firm understanding of the business is a good beginning.

A few comments about worker skills are in order here. Think about the people in the organization where you work. Now think about what each person's job is *supposed* to be—perhaps they are computer programmers or quality professionals. Finally, think about how well the skills required to perform each job match the skills that the person has who is performing the job. Are there any gaps? If so, how big are they? We believe that a large percentage of American workers are overmatched with the expected demands of their position. This is one of the predominant reasons for organizational infighting, CYA, and status checking. It is not that people do not want to perform value-added work; it is simply that they do not know what to do. That is why there are reams of organizations whose names indicate "planning" or "management." Even within the value-chain, many workers have not been adequately trained to perform the functions they need to perform. They may have been cast to the wolves with little or no formal training. The lesson is that people need skills—*hard* skills. Companies need computer programmers who understand the languages they are asked to program in; they need technicians who are trained on the piece of equipment that they are expected to service; they need managers that know more than schedule checking and have a clue as to what it means to do the work of their subordinates. We cannot begin to address the areas that skills need to be developed in for a given company. A lack of skilled workers at all levels in an organization will make reengineering a failure in either the transition period or at some later time. Of course, the skills inventory discussed previously will give you ample warning about the situation that your business is in.

Another problem is what we call the "technology abyss." Over the last decade, technology advances in the computing industry have wreaked havoc on the capability of the workforce. A classic illustration of this problem is the overwhelming number of data processing professionals who are untrained and unskilled in state-of-the-art computer system development environments such as distributed computing. We have seen companies that are considered high-tech where a significant percentage of the workforce lacks an understanding of how the high-tech products work. Today, almost all jobs either use technology or impact its creation. Order clerks use systems. Market researchers create information used in decisions about the next product. Many companies will have a workforce that is "technology illiterate" as they embark on reengineering projects. Every worker needs to be educated in the appropriate technology for a company to flourish.

As Peter Drucker points out in *Post-capitalist Society*, the age of the "knowledge worker" is here. Hard-working, empowered people will be effective only if they master the skills and possess the knowledge that the job requires. Mechanics, order clerks, and senior management all need to retrain in those areas that have radically changed their professions. A clear understanding of the business that the company is in, coupled with an emphasis on understanding how technology works, is a starting point for success.

Senge/Covey Self-Mastery

It is time to go from the hard to the soft. Of course, employees must have the appropriate skills to perform their jobs, but equally important is their mental health. There is a strong link between good mental health and superior job performance. In some respect, the entire quality improvement decade was geared to help the psychology of the American workforce. Quality improvement teams and worker empowerment programs aim to make workers feel better about their jobs (the "make them feel good and they will do good" technique). We can all relate to failures that have occurred with these programs. It is reasonable to say that the quality movement has left a large number of allegedly "empowered" workers who are extremely unmotivated. To a large extent this is so because many managers never made the transition from the old way to the new way. Before a company can take the team organizational approach that reengineering redesigns will demand, it must spend time teaching people to master themselves. Two authors who are especially good at conveying these concepts and their importance are Peter Senge and Stephen Covey.

Both Covey and Senge offer powerful suggestions: the key to organizational success is the self-mastery of the people in the organization. Their approaches differ in that Covey offers more of a "self-help" approach (at least in written materials). A realistic picture of where almost everyone stands in terms of personal mastery is gained by reading Covey's *Seven Habits of Highly Successful People*. Most of its readers whom we have spoken to have said that they have light years to go before they will *live* the seven habits. Taking the seven habits as a baseline for people who have mastered themselves, it is hard to imagine effective teams of people whose members do not live within the framework of the seven habits (comparing them, for example, to a hypocritical and ineffective mass of people who follow a religion and live none of the principles). Team discipline generally breaks down because of a lack of self-discipline. Unfortunately, as tempting as it may be to insist on self-mastery training for all, mandatory enrollment in such programs probably would backfire.

One of the problems with trying to institute self-mastery/self-help programs is well characterized by Senge. Although he realizes the importance of people creating their own visions, vigorously seeking knowledge, and building the creative tensions that are vital for the organizations where they work to flourish, he also warns against the coercive effect of making people attend personal mastery courses. Senge's alternative is that people can achieve self-mastery through osmosis within a "learning organization." We believe that this is too lofty an ideal; rather, we have seen that explicitly making people aware of these programs can be especially powerful and contagious. Helping people understand what the programs offer probably will spark enough of a grass roots effort that a fair number of employees will voluntarily seek this training.

Team Building

Almost all organizational designs of the nineties stress a team dynamic. Our experience is that "teams" work well as an operational result of reengineering (what the workforce will look like) and in the implementation of the reengineering itself (the development of systems, for example). Even in situations where the process can be completed by a single case worker, the case worker usually is part of a team with other case workers and management. Therefore, building a team properly is very important.

It is a big mistake just to form a multifunctional group

of employees, organizationally aligning them, putting them in the same room, declaring them a team, and calling it a day. That approach is a certain recipe for disaster. Although this point may appear to be pure common sense, it is something we have seen management overlook countless times. It appears that managers have forgotten their childhood. Did all the kids in your neighborhood play together in harmony? Can anyone really expect that just because we are all adults in a work environment that we can work together with no problems? This line of thinking is ridiculous. Teams must be built carefully and then guided to work together in a manner that builds rapport.

Many consulting firms and human resource groups recommend using personality/psychological profiles such as Myers-Briggs to assist in team formation (Note: these tests also can be used in job matching). Admittedly, there is no alchemic mix for forming the correct team. However, Myers-Briggs and similar profiles are very helpful in avoiding disasters. They are particularly geared to point out oil-and-water type differences in people. For example, if every team member is a disorganized, outgoing extrovert, the team may have a problem performing a heart transplant. At the other extreme, avoid using these tests as a mechanism to build a perfect team; no test is nearly that precise. Instead, use these tests judiciously. The results also are extremely valuable when teams are sent off to team-building sessions. The team can strengthen itself through the differences in people, as members really try to understand one another better.

Finally, a prospective team should be sent to group training. Bell Atlantic has been utilizing the services of Zenger-Miller for team-building training after the implementation of a reengineering project. Their experience is that it helps individuals make the transition from old paradigms to new ones. Many businesses need assistance to break out of old paradigms of strict management, control, and self-protection to a new paradigm that stresses teamwork, empowerment, and looking out for one another. They must understand that the strengths of their organizations are at all levels of the pyramid in Figure 17-3: subject matter knowledge, self-mastery, and team building.

Attacking the Specific Skills Needed for a Reengineering Project

Organizations will succeed only when they take an organizational development approach as outlined above. In a reengineering project, several more specific training areas must be addressed. To a large extent, the training areas discussed below are "normal" for introducing new computer systems or procedures into an organization. It is important, however, to consider these areas in terms of the radical shift that will take place with the process redesign.

Systems Training
If the reengineering solution requires the implementation of new information systems, it is obvious that employees must be educated on how to use the system. Besides training employees on the functionality of the new systems, organizations often have to overcome a lack of expertise in using the new technologies that are being introduced. For example, most user interaction in today's systems is done via "mouse-driven" user interfaces as compared to the older terminals where function keys were state-of-the-art. This usually does not cause overwhelming problems, but having a workforce that understands the basics of windowing applications is a plus. One thing that works well is giving people computer games that require the use of a mouse. This operates as a positive reinforcement for using the new tools.

Process Training

The need for process training is as obvious as the need for system training. It is important, however, to remember that you absolutely do not want to give detailed process training until the workers acquire basic skills and understanding. To best accomplish process training, it is important to consider the general concept of systems thinking. Teaching employees to learn the concepts articulated by Senge may be the single most important educational investment that an organization can make. Having the employees think beyond a linear process to a more holistic interdependent process will ensure future improvement and lessen the chances of re-forming "smokestack" functional organizations. Nevertheless, the overall goal of the process training is for all people working within the process to understand their role, how it fits into the overall process, and how their process fits into the company—its goals and its customers.

Facilitation/Quality Management

Over the last couple of years, we have personally experienced the benefits of learning and using group facilitation skills. These skills have been invaluable in a variety of ways. For instance, facilitation skills enable groups to have effective meetings where work actually gets accomplished. Before using these techniques, we witnessed the problems that plague many meetings: useless dialogue, side conversations, and zero productive output. In addition, facilitation skills are essential for coaching newly formed teams of people. As outlined in Chapter 16, management must learn to facilitate and coach teams for the team dynamic to be successful. What we are discussing is a critical paradigm shift. Although a large portion of the challenge is to overcome the psychological or "mind-set" paradigm, some skills must be taught to make facilitators effective. Here are a few examples:

1. *Listening:* Taking the time out to really listen to other people's problems (teams always have problems) will help the team grow. As Covey points out in *Seven Habits of Highly Successful People*, being able to listen objectively without bringing in your own biases is a hard skill to master.

2. *Allowing for mistakes:* Teams inevitably make mistakes. Managers are capable of making some decisions in an instant because of their experience; but as long as a manager is in the loop, the team is not being empowered. A manager in the new environment must learn to rally the troops after failures and quickly reward successes. To quote one manager, "You may need to have your tongue replaced from biting on it so many times, but every time you have a need to bite on it progress is being made."

3. *Negotiation:* Handling disputes within the team and between the team and management requires good negotiation skills. Disputes will be resolved in a shorter period of time once trust is built between members of the team. Making both sides walk away from a dispute feeling they have learned something, and without feeling they have lost the fight, is an important part of building trust.

The strategy for accomplishing the change to the new management style is a process in itself. In the Curriculum Strategies section, we discuss some curriculum approaches that help accomplish this goal.

Statistical Quality Control (SQC)
At the foundation of the simulation models discussed throughout this book are statistical concepts. By induction then, in order to continue to use these modeling tools once the reengineering team walks away, there must be an investment in educating workers on the tools themselves and, more important, in SQC. In Chapter 16, we high-

lighted the role of modeling the new organizational paradigm. In practical terms, simple concepts such as variation are extremely important to all levels of the new organization. In any type of operational process, workers need to be able to read and understand quality tools such as control charts, run charts, histograms, and so on. To improve a process, one first must understand it.

CURRICULUM STRATEGIES

The education plan will succeed only if a curriculum is designed to educate workers in all the topics that are discussed in this chapter. Most colleges have initial core requirements for all students. Then they concentrate on one subject or "major." As we discussed earlier, a similar model should be used in retraining the organization. For example, the workforce first should be trained on the fundamentals of organizational development: the areas identified as subject matter knowledge, personal mastery, and team building. Then, more focused classes should take place as the time to cut over to the new processes comes closer. These programs should focus on the details of how the process will operate: process training, system training, and quality techniques.

The most successful corporate training curriculums do not exclusively utilize classroom settings for educating workers. Most of the "work" that occurs in businesses does not easily transfer from the classroom to where the work gets done. Classroom training falls short because it is not context-sensitive. The majority of the things that people need to learn to perform new job functions can be taught only by actually performing the functions. The remainder of this section discusses some techniques that have been utilized successfully. By no means is this the end of road for learning how to build education strategies. We encourage you to tailor a program that will meet your needs. We recommend, however, that you try not to rely solely on external sources to deliver the training. Much can be accomplished by the reengineering team's use of some of the concepts discussed in the remainder of this chapter.

Using the Lab Environment

As described in Chapter 9, the redesign process takes place in a flexible laboratory environment, free of rigid organizational boundaries and administrative policies. You should exploit the lab environment as much as possible for all activities. As it is being used for the analysis, redesign, and system prototypes, it is only natural that it be used as a place to conduct training. There is no better place to conduct process and system training than the place where both are being used. Unfortunately, this is easier said than done, especially for the initial release of the system and the process. Many activities are performed in the lab environment, so it will be very difficult to manage training in the same physical space (note: be careful not to make the lab a bottleneck in your reengineering process). Shuffling people in and out of the space for training purposes can be particularly chaotic.

There are a variety of alternatives to forcing people to stomp on each other in a crowded lab environment. The first and the simplest is to schedule training in the lab environment during nonbusiness hours. The second alternative is actually a systemic approach to using the lab environment: instead of using the same employees in the lab for the entire length of the project, design a program to rotate a large number of the workforce through the lab throughout the entire project. There are tremendous bene-fits to this approach:

1.   More employees get involved in the project.

2.   Input from more workers is available.

3.   Each person who revolves out of the lab environment is another potential PR person for the project (this can backfire if people have negative experiences).

4.   There is less need at the end to retrain people from scratch. A refresher may be adequate.

The largest downside to this approach is that there may be a loss of continuity in the process redesign phase. For instance, if the lab environment also is being utilized as the breeding ground for a self-directed team, it does not take an Einstein to realize that it is difficult to build a team when management is playing musical members.

Microworlds and Scenario-Based Training

Two tightly linked concepts are the idea of microworlds and scenario-based training. To really understand micro-worlds, we suggest that you consult one of our favorite references, Senge's *The Fifth Discipline*. The skinny version goes something like this: humans learn by doing, but they also learn best while playing; the concept behind micro-worlds is that simulations of the "real world" are created so that people can learn by doing examples of likely real-world scenarios. In one of the examples in *The Fifth Discipline,* Senge described a microworld where corporate leaders were going to discover what would happen if they tried to reach a strategic goal that they had set earlier. While laying out the plan for accomplishing the goal, the executives learned about one another as well as about how the plan was flawed. The beauty, of course, is that this was done on a two-day retreat, so no real harm was done. Senge gives compelling reasons why microworlds will be the learning tool for the future. The possibilities do seem endless considering the commercial emergence of technologies such as virtual reality. Simulations of possible real-world business environments are just around the corner.

This begs the question; that is, how will microworlds help on a reengineering project? For one, they may be used by the reengineering team itself to challenge some of the assumptions made in the process redesign that do not lend themselves to computer simulation. For training those workers impacted by the redesign, a variety of microworlds can be set up to try out the new organizational relationships. For instance, role playing between management and nonmanagement in a microworld setting can be extremely useful in sorting out potential confrontations. If the process requires groups in different locations to interact, the creation of a microworld to discuss group interactions in some real-world scenarios is invaluable to the team and to problem prevention.

Somewhat more practical than microworlds is the concept of scenario-based training. In some sense microworlds create scenarios, but scenario-based training is the "real thing." Remember, a reengineering solution involves an overhaul to the process, and in many cases the information technology changes. The process redesign always has some scenarios inherently built into it. If there is a new way to take customer orders, create scenarios that are likely to happen. Use the new process and systems. Inject likely errors. Let people learn the process by doing it with the real tools. We have seen this work well in the introduction of new computer systems for new services, where the system development lab was the playing field for the scenario-based training. The users of the system were brought to the lab for training, and the scenarios were played out using the real system in combination with computer programs created to test the software. As the programs were designed for both failure and success scenarios, the benefit of a "play out" with the eventual users of the system was gained at no additional cost. With a little ingenuity, scenario-based training can be used on any process. Take an inventory of the tasks and the tools that will be created to make reengineering happen, as they become reusable assets similar to the test tools mentioned above.

On-line Methods/Human Systems Interaction

Although the computer system application being created to support the reengineered process should not be expected to be a training tool in itself, there are advantages to using the system to fill in the gaps as well as to allow users to learn as they go. One of the ideas discussed in Chapter 12 is the use of work flow software. The process is available on-line, and the system both keeps track of and initiates process steps. Each step in the process that the system knows about boils down to a task that either the system or a person will execute. The system can store and display vital information to the user for each of these steps.

Many times millions of dollars are spent developing what are called "Methods and Procedures" or "Operating Procedures," which are designed to be job aids for employees working within a process. These volumes of "how to" instructions are cumbersome to use at best, so even the most comprehensive and accurate procedures manuals go unused (also many trees are killed to produce them). We propose that information about each process step be kept in a text file that is readily available for display whenever a user is executing a particular task. In practical terms, this function is similar to context-sensitive help that is available via hypertext in many PC applications. The difference here is that the "help" available for the process step goes beyond telling the user how to kick the system to get it to do what is needed to continue; it also gives vital information about the process step. The text should include information on what the task is, what its main purpose is, how it enables other work to be executed, and how it fits into the global process. Also, this method is far superior for updating procedures. Instead of distributing new manuals to everyone who had the old one to ensure that everyone has the most current version, changes need to be made to the system only.

The use of "on-line methods" is a gap filler to augment a comprehensive training program. However, the use of such capabilities within a system opens up a variety of continuous learning mechanisms. In essence, by employing "on-line methods," a dedicated effort has been made to expand the horizon of the human–system interaction. In developing "on-line" methods, time should be spent in figuring out how to link users together in some meaningful way, beyond the work flow capabilities, so that information can be shared about the process. As there are many ways to do this, we provide a few brief examples here. The users should be able to add text about each process step so that other users requesting help for that task can get some up-to-date real-world experience from other users. This information will be invaluable to a process improvement effort. Another example is the use of electronic mail features that can help users in different locations to communicate. The electronic mail capabilities should be part and parcel of the application so that users are not expected to log off, go to their e-mail system, send e-mail, and then log back onto the application.

To start thinking about how to make the computer

application work in ways that support continuous improvement, we suggest that you spend some time learning about human–computer interactions. A good source to start with is *User-Centered System Design,* a compilation of articles edited by Donald Norman and Stephen Draper. Two articles in the book are particularly relevant to the ideas mentioned here. One of them, "Helping Users Help Themselves" by Claire O'Malley, discusses ideas that are critical to on-line procedures. The other article, "Helping Users Help Each Other" by Liam Bannon, presents concepts related to using the computer in ways that assist the continuous learning objective.

Structured OJT

There has been a lot of recent criticism of "on the job training." The biggest problem cited is the wide fluctuation of results that occurs with its use. Among the reasons for wide variation are:

1. *Unstructured programs:* When a worker is asked to show a new person "the ropes," the last thing in that worker's mind is "How did I train the last new person?" or "How do the other people do the training?" Chances are that no two training sessions will be the same.

2. *Bad habits:* Very often there are no standardized ways to do a task. Instead, workers develop their own methods of accomplishing tasks, including making decisions about what is important and what is superfluous. These "habits" are passed on during on the job training.

3. *"Virgins in the volcano":* When new workers are sent to the front lines to learn how to do a job, they are sent out with very little introduction to how the company operates or what their job is really about. This frequently happens today with the increasing presence of temporary employees in the workforce. A new employee may never get assimilated into the environment and may spend several months just trying to get his or her bearings.

4. *Experience:* Most employees have no experience in formally presenting courses. This lack of experience tends to make the information presented come out randomly, based only on the small number of permutations that arise as the OJT is occurring. For instance, if a worst-case scenario is presented to the worker during the OJT sessions, the new employee may think it is the usual case and be confused when confronted with normal routines as they arise.

Given the track record of OJT, it may be surprising that we recommend it. However, we believe that there is no better base of knowledge than what exists at the level where the work is being performed. Also, reengineering demands flattened organizational structures, so the future of training in the workplace may depend on doing OJT correctly. The goal then is to harness this knowledge and make it work for you. Really, it is as simple as attacking the problems associated with the typical OJT. Here are a few ideas.

1. *Train the trainer:* Not all people will be good candidates for trainer. Use the skills inventory to pick the few who will, and get some professionals to assist them in developing a curriculum. Having a designed curriculum will help workers with experience, bad habits, and structure problems.

2. *Use designed Feedback:* Make sure there is a formal evaluation of the training regardless of how informal it is. In addition to formal evaluations, follow up to determine how proficient the employees who were trained on the

subject are at performing it in the work environment. How can a system be improved without any data?

3. *Make OJT part of the business strategy:* Make OJT part of the way of doing business, not just a way to fill in the gaps. You should have trainers available for all the functions and ensure that the time spent in OJT is built into their schedules.

4. *Use other methods too:* OJT only can fill a small gap in an overall curriculum. If employees are never assimilated into the job so that they understand the big picture of what the process is about, they will still have several months of catch-up time, or worse yet may never feel as if they "fit in."

Benefits of This Approach

1. *Employee participation:* In addition to some employees' learning "normal" functions, they will have other challenging opportunities.

2. *Gaining time:* Scheduling professional trainers over a long period of time can be quite difficult. By having employees capable of training others, more training can be done at once.

3. *Eliminating project management headaches:* With the complexity of rolling out a reengineering solution, eliminating some dependencies on outside resources simplifies the overall process. Additionally, this strategy can help cut expenses.

THINGS TO AVOID

1. *Expecting a magic formula:* There are no magic formulas to train the workforce resulting from reengineering. Fortunately, in many respects the "right" things to do are the same as what may work well when a new system or product is developed. The critical difference is that there are more changes associated with reengineering. Know the climate, know the corporate capabilities, and make prudent choices.

2. *Forgetting self-mastery and skills:* We cannot stress enough that at the core of all corporate capabilities are skills and personal competence. We have seen organizations latch onto team building, hugging and kissing, and empowerment programs as a way to make things better. In most cases, these organizations have failed miserably. They did not realize that the people had no personal ambition, and the skill sets were a fundamental mismatch with the task to be done. Do not fall victim to this phenomenon. Make sure that the necessary skills will be available to get the job done, and if not, change the design or get new people.

3. *Keeping people because they are all you have:* Be prepared to change staff. Some employees will not be able to make the transformation. Either put them to the side out of harm's way or remove them from the process entirely. This may sound cruel, but it is not nearly so cruel as bankruptcy.

18

Retool

Even though the new business design works well in the laboratory environment, the real benefits of the project will come only when the new design makes the transition to the other organizations. In taking the new business design from the laboratory environment to the rest of the organizations, some of the most severe complications involve the technological aspects of the project. The wide-scale cutover to a new information system must be planned prudently in order to avoid risk and increase the chances for overall success. Yet, organizing the other organizations into case teams like those in the laboratory also presents numerous problems. This chapter describes the many problems that can arise during the transition and explores the advantages and disadvantages of different transition strategies.

TRANSITIONING NEW TECHNOLOGY

It is not possible to list below all of the system-related activities that must take place in order to smoothly transition new technologies. (For example, we do not discuss all the perils of the installation of new system applications that may arise.) Rather, we focus on some major concerns that are especially acute in reengineering. The system delivery process required to support reengineering needs to be different from the normal deployment methods used by the information systems group. Just as the process for systems development must be replaced by rapid prototyping and application development tools, so must the process for system deployment be restructured.

In the previous two chapters, we discussed how the reengineering lab can be used as a catalyst for reorganization and retraining. No matter how the lab is used for the human element, there are significant issues relating to its use in system introduction. These issues exist no matter whether the system functionality has been prototyped and evolved. Making a system operational for everyone has repercussions that occur only when a full-scale deployment is attempted.

"Rightsizing" is a term used to describe the re-hosting of applications from mainframes to smaller, distributed platforms. Most reengineering projects usually involve some sort of rightsizing. This occurs because a process-driven business is always a good candidate for client-server architectures (as explained in Chapter 12). Unfortunately, most documentation in this area focuses on the information technology problem (hardware, software, etc.). Therefore, we will try to look at the problem from a process perspective by discussing alternative cutover strategies and their potential positive and negative impacts on the introduction of the new technology.

There are a variety of permutations that a cutover

or transition strategy can take, but there are two main approaches: the controlled introduction and the flash cut. In a controlled introduction, the new system is gradually made operational by first introducing it to a control group. With a flash cut, the system is rolled out to all users over a very short period of time. For our purposes, because we recommend a reengineering strategy that actively includes prototypes and system evolution, any cutover strategy that goes from the small work team to all the users in one fell swoop also will be considered a flash cut.

Controlled Introduction

Ensuring that the system is operational for a segment of the business is the rationale for a controlled introduction. Hopefully, you are by now convinced that the evolutionary approach to process redesign and system development is the best way to accomplish reengineering. Many of the same arguments apply for the incremental transitioning of new technology. The following sections describe the advantages and disadvantages of this approach.

Potential Advantages
A major advantage of the controlled introduction approach is that the "production" hardware and software have already been soaked in the "real" environment. What does this mean? Many evolutionary development processes rely on the use of a development environment for the testing and the debugging of each iteration. We strongly recommend that the production platform be used as much as possible. There are two main reasons for this:

1. Reengineering projects present excellent opportunities for rearchitecting or rightsizing the computer centers. The evolution of hardware platforms from mainframes to minicomputers does not necessarily come easily. Therefore, having the hardware work in the field many months before *all* the users must use it allows enough time to sort out the usual deployment problems (e.g., "missing" commercial software packages, flaky communications lines, etc.). The bonus, however, is that the systems support personnel get to touch and feel the application and platform long before they must. Reengineering projects typically require rearchitecting so that support personnel may not be used to the new architecture (e.g., client-server). Thus, getting these people up to speed early can help.

2.   An enabling process exists in the software delivery process. This process includes activities such as software installation and version control. Ironing out the bugs in the process will reduce errors at the end and minimize finger pointing.

Another advantage of a controlled introduction is that it lowers risk. Whenever something can be broken into manageable pieces, risk is reduced. The reason for this is that less change is being introduced at one time, and the opportunity to "back out" is more likely to be available. As we will discuss later in this chapter, however, finding manageable pieces is not always easy.

Potential Disadvantages
With a controlled introduction, information systems may retain duplicate functions for a period of time. It is not very often that the process and the system are so new that there is no relationship between what the new process does and similar work in the old environment. For example, if a claims process for an insurance company has been extended to be performed by the local agent rather than central claims adjusters, the process has been radically

redesigned, but the "job" of claims has not simply disappeared. Therefore, the system that supported the old environment must run in parallel with the new system for a period of time. With such redundancy come both complexity and cost. Managing two separate processes so that the result appears seamless to the customer can be quite a burden. Maintaining two information systems with the same basic functionality can be expensive if the cutover is extended for a long period of time.

In a phased approach, computer applications that support similar business functions inevitably must exist side-by-side. In reengineering, this can be especially problematic because the redesigned information systems generally replace what five or six systems used to do. The coexistence of computer systems that support different business processes (and have different hardware and software needs) is something that any information systems group would like to avoid. Nevertheless, it is the responsibility of the reengineering team to evaluate the pros and cons of this approach.

To illustrate some of the challenges of a controlled introduction, we offer an example. Suppose that you have a sales support system that contains information about customers of your telecommunications company. The salespeople use the information to assist in the sales and marketing functions that they perform. When a customer places an order for equipment, the salesperson enters the order in the sales support system, which in turn, spits out a report. The report then gets faxed to the factory for further processing (and data entry in its own systems). To eliminate islands of information, the reengineering team decides to provide real-time inventory information so that the order process can be more customer-driven. This requires the development of a new system that replaces the sales support, the inventory, and several other ancillary systems. The reengineers successfully try the new system with a single major customer and one sales group.

Now, it is time for a wide-scale cutover of the system. However, several major customers in the eastern part of the country are in a panic mode. They are depending on the company to fill the orders that were placed recently as well as the next few months' orders. Thus, the reengineering team decides that it is too risky to cut over the eastern region at this time, and they choose a phased approach that firsts transitions the western region to the new system and process.

While this decision averts potential customer service problems, the reengineering team and the information systems personnel now must deal with difficult technical problems. At this point, the eastern part of the country is still using the old systems and old process, and the western part is using the new system and new process. What about the data on those customers that have *both* East and West Coast facilities? For example, one customer has a single principal ordering agent for all sections of the country. What if that agent gets a new telephone number, or what if a new agent is hired? How do the systems (old and new) both maintain the agent data? Such redundant data are a source of errors. Also, synchronizing redundant data across systems can drain the IT resources of the project.

Strategies for Controlled Introduction

It is easier to segment a phased approach by *customers* rather than by product, service, or geography across customers. Let us revisit the data-sharing problem of the telecommunications company. The problem of which system to use to update the customer information and how the two information sources were to be kept in synchronization occurred because the phased approach was based on geography. Both sides of the country needed to access and update the customer information, and both systems were required to process customer orders for that customer. If a phased approach based on customers were used, the problem with the system would not exist. From a data perspective, the systems could function more independently.

Flash Cut

Reengineering sometimes is said to be an all-or-nothing proposition. For those with the gung-ho attitude that "we have nothing to lose," a flash cut implementation of the system and process may be just right. It eliminates the need for a hybrid environment imposed by a controlled introduction. Also, it quickly vaults the company into the new business environment, allowing it to realize the reengineering benefits that much sooner. However, high risk is the primary reason for *not* flash cutting. No matter what the deployment effort may entail, the last thing that people want to do is to use a backout strategy. In reengineering, the changes usually do not naturally lend themselves to an elegant backout. For example, backing out of data conversion can get messy when a lot of time has been sunk into the conversion itself. It may be impossible to restore the old environment fast enough to avoid affecting customers. Another problem lies in making sure everyone is on the same page come flash cut time. When new processes are being put in place, special precautions must be taken to ensure that people know what is going on before it happens. In general, a lack of experience with new architectures and processes makes reengineering projects especially vulnerable to the problems that can arise during a flash cut.

A flash cut would be appropriate in certain scenarios, such as the following:

- The scope of the reengineering project is small.
- There are overwhelming coexistence problems.
- There are few new users.
- Customer disruption is unlikely.

Small Scope

A project with a small scope, such as a subprocess, may provide a reasonable gain without the large risk associated with transitioning large projects using a flash cut.

Overwhelming Coexistence Problems

Sometimes it is just too difficult to come up with a viable phased approach. Therefore, it is important to start thinking about the transition to the new environment as soon as you define the scope, and certainly during the redesign.

Few New Users

If a small number of people are impacted by the change, that increases the odds of a successful flash cut.

Customer Disruption Unlikely

The transitioning of redesigned external processes (i.e., those involving the customer) usually affects the customer in some potentially negative way. Entirely internal processes, of which there are few (one example is a travel reimbursement/business expense process), can be transitioned via a flash cut because the customer is unlikely to be impacted.

Data Conversion

A significant challenge exists to convert the data that supported the old business process to the representation of that data in the new business process. A maxim in reengineering is to "rethink" processes, but unless you plan to forget who the customers are or last year's sales volumes, you cannot eliminate existing data that supported the old processes. The seemingly simple task of getting data from the old systems and putting it in the new system is hardly mundane in reengineering projects.

Speaking of data, one thing is probably a given as a result of reengineering: a bunch of data should no longer exist. On most reengineering projects the data elements necessary to make the process work are usually 30 to 40 percent less than what was needed to support the old process. There are several reasons for this, including the elimination of administrative data that supported the "make work" management jobs as well as the reams of tracking information used to tie together the disjoint process. Having fewer data to worry about makes the job of conversion a bit easier.

The same thing cannot be said for refining the data. At the crux of the process analysis is the effort to design a new logical information model. In the logical model, the data are represented at a "normal" form level. Unfortunately, the old systems probably did not have this "normalized" view of the data in their physical databases. In effect, what you end up with is some data elements in the old system that now represent more than one data element in the new system.

Most reengineering projects have to resolve poor data integrity. Processes that are wildly out of control usually have many points of rework and desynchronization, and with them comes poor data integrity. A broken process can cause the telecommunications equipment vendors never really to know for sure what the customers installed in their office. There may be no place in the old process for the information to be updated at installation time. Therefore, mismatches exist. We can almost guarantee that one or more of the existing systems will contain incorrect data. It is the job of the reengineering team to deal with the problem, and we have a few suggestions to help them get started.

In the New York metropolitan area, there is a department store called "Fortunoff's," which calls itself "The Source" because of the wide assortment of goods it sells. Our approach for locating the data that you want for the data conversion is also called "the source": the goal is to go to the original source of data and work at validating the data *at that point.* Process boundaries are often also sources of information, the most common example being the customer's input. This is the best place to validate the data to be used in the new process. For example, if a customer has your product at his or her location, there is no better way to get good data than to go there and look at it (e.g., physically scan a barcode on the equipment). This may prove better than any copy of the data found in any of the systems. A secondary choice will be the system closest to the source (the customer). The more validations the system has and the fewer hand-offs of data before they are entered in the system, the better the data are. Other boundary points to examine include the outputs of the process. For example, a billing system serves as an output for many extended processes. These repositories also may help in providing the most accurate data for the new process. Customers complain when they are billed improperly, causing greater attention to be paid to the billing system's data integrity. (Some billing systems also act as inventory stores. This is generally a bad idea but may be useful in a data-conversion effort.)

A brief example from a telecommunications company illustrates the uses of sources and outputs in performing data conversion. To make telecommunications networks work, the company needs to have information about the customer's equipment to get the service up and running. This information somehow must be downloaded into the network so that customer phone calls can go through. Matching information also must find its way into the billing system so that when phone calls are made, they are billed to the correct customer. Using the guidelines described above, we went to the sources when we found problems in synchronizing the customer data in network and billing systems. The sources here are the customer location (which drove our process to begin with), the network (the meat of this particular process), and the billing system (how we are billing the products and the services). Looking at these sources told us much more than looking at the intermediate systems. In converting data, we were able to identify which of the sources was most accurate and to use that source as a starting point for the data in the new system. Also, we were able to identify discrepancies *between* sources and worked to resolve them prior to the wide-scale cutover.

NONSYSTEMS ISSUES

More problematic than systems issues relating to transition are the process implementation and its effect on employees. The process implementation itself is easier than the human problem because it almost necessarily dovetails with the systems approach (i.e., the process is transitioned either by a flash cut or by a controlled introduction). The people concerns are much more emotional than this and require good planning. The goal of this section is to get you on board with the implementation of the process and also to raise your awareness of the people issues involved in transitions.

Fundamental Process Rollout Strategy

Every new process implementation that we have been involved with is surrounded by confusion, frustration, and sometimes panic. Even if people know that changes are going to occur, there is still the potential for problems when the changes actually take place. Further, no matter how well things are planned, once people start moving to new locations, using new systems, and performing different responsibilities on a broad scale, there is a tendency toward chaos. We believe that the best transition strategy minimizes interference in the overall environment. To achieve this, we recommend that the strategy for rolling out the process be worked part and parcel with the strategy for rolling out the system. Keeping in mind some of the suggestions for system rollout strategies, think of how the process fits in with them.

If you consider the multifunctional case teams discussed earlier, a good strategy for implementing the process is to do it one or perhaps two groups at a time over the course of several weeks. The systems rollout that calls for a phased approach by a customer is applicable here. As the overall strategy may be to cut over the systems over a period of months, the process introduction should parallel those time frames. Depending on the robustness of the data conversion capabilities, the actual rollout of customers to work groups can be broken down in more detail. The rollout should work much like a spigot that the reengineering team controls: the more it is open, the more customers and hence work teams are cut over to the new process.

The alternative way to roll out a process is to flash cut. The risks associated with the system aspect of a flash cut usually are significant enough to steer people away. If not, the problems that are likely to arise on the process end may be the last straw. The assertion that all the planning in the world does not make transitions go smoothly is applicable in spades when a flash cut is attempted. Common sense (and certainly the law of probabilities) says that if you try to change many things at once, there is a good chance that not all outcomes could have been determined beforehand. First, people usually have enough trouble with the paradigm shift that has taken place. Getting used to a new way of doing things is not easy. Second, dealing with inconsistencies and unexpected problems as they are occurring on several fronts can be overwhelming. The part of the reengineering team responsible for transition will be overwhelmed if problems are sprouting up all over the place, and the team will become ineffective. Employees working in the process will start the rumor that "everything is broken," and recovering from this cynicism will be difficult.

The People Side

The goal of this section is not to spell out all the overwhelming psychological implications of major change. Horror stories have been well documented, and managers should all be aware of them by now. Nevertheless, we can offer insights into some of the practical people problems that should be considered. Our emphasis is on identifying potential problems and describing the programs that companies have used to deal with them.

Upfront, you must account for the cost of moving people and planning for floor space changes. These seemingly small administrative issues become show-stoppers at worst, and at best are annoying problems that can slow down the implementation. Budgets for moving people to new cities or even new buildings are almost never considered when a project begins. Floor space planning in major corporations can be a complete nightmare. Talk about turf problems—this *is* real turf, and many people do not want to give it up.

Forcing the Troops to Move

Many new organizational designs call for the co-location of workers performing the various steps in a business process. When employees must be co-located, there is an underlying assumption that they are not all currently in the same place. So, someone has to move. The movement can be to different buildings in the same city or to different cities altogether. Depending on the situation, different problems come up, but in either case some problems definitely will arise.

Despite the commonplace nature of business-related relocations, such a move often is traumatic, especially in traditional companies. In these companies (e.g., steel, autos, telecommunications), where the workforce may be non-management, unionized, and older, moving does not come naturally. The workers face a variety of problems, ranging from the low prices of sluggish real estate markets to the inability to leave family behind.

The mechanisms for dealing with these problems are as varied as the range of problems themselves. One company that was moving a largely nonmanagement workforce from all over the country to a small town had the problem that many of its employees had lived in their home towns all their lives and did not know how to sell their houses. The company hired a real estate agency to educate the employees on the home buying/selling process, and later it became the listing agency. This novel approach was fairly simple to implement, but it takes an aware management to foresee this problem.

Great problems arise when people become depressed after leaving a place that they lived in all their lives. Sometimes the severing of deep roots causes alcohol and drug problems. If such a situation seems likely, or maybe just in case, hire some employee counselors to assist people with the transition. Catch these problems before they have a chance to develop. By the way, do not misread what is a common situation with some workers. A certain percentage of the workforce may reluctantly agree to move, resent the company for it, and subsequently do absolutely no work. Remove these people, or at least make sure that they do not poison the environment.

Of course, it is much easier to move people from one building to another within the same city than to move them great distances. Other than the normal backlash that is expected when several changes occur at once, the only problem that we have seen is that some employees resent no longer working with old friends and the disbanding of their coffee klatches. We suggest that you give employees time to get together with their old colleagues. People have a difficult time maintaining friendships, and the company should try to encourage them whenever possible. In this case, it will be in the company's best interest to have employees from various locations keep up with the activities of other organizations in other locations. The employees will become better versed in company affairs in this way.

Unfriendly Innovation

Technology may be an enabler, but it does not always come naturally. Reengineering solutions that utilize technology enhancements should allow some time for workers to adjust to them. Obviously, the greater the leap is, the greater the adjustment. For example, you cannot expect someone who programs by using a card reader to become literate in a high-powered workstation overnight; or you cannot wonder why the marketers who have been involved in highly manual processes are taking so long to adjust to the groupware software that links them with engineers. Just because we are in the 1990s, it does not follow that everyone is literate on all technology platforms. (Most of us have had to struggle to figure out the VCR!)

Patience is definitely a virtue in this case. Give the employees several months to become acclimated to the new tools. Just knowing that this learning curve exists is a plus, but staying patient is not always easy. Here is a small example of what can happen if you get too antsy. A manufacturer of Automated Call Distributors (ACDs) decided that its customer support staff, which previously was divided on a hardware versus software boundary, was going to have each person be responsible for both hardware and software. This may have made sense, but the company's approach did not work very well. One of the best software people that the company had was scared half to death of the rigorous hardware training course that took place over a week. By the end of the course, not only was the employee incapable of doing the hardware tasks; he also had lost self-confidence so that he started to seriously doubt his ability. This type of result must be avoided.

THINGS TO AVOID

Most managers do not care much about the little things, but a successful transition to a new work environment can be successful only if all the little things work. We do not suggest that one person has to do everything or that the lead executive know that the employees are using computer games to learn mouse-based user interfaces. However, all levels of management must be aware that the little things are out there. In addition, there may be a tendency to think a problem is solved when the prototype works with a few customers. A successful prototype tells you that an idea *can* work, but it is up to the team to make sure it *does* work.

19

The Next Wave

This book was designed as a guide for successfully implementing comprehensive change *in 1994*. We are confident that the implementation approach we have described will be applicable for quite some time. The concurrent examination of processes, organizations, and systems will continue to provide a better understanding of a business's behavior and its problems. Further, an evolutionary implementation of the new business environment will continue to provide benefits more quickly while reducing a project's risk. Yet, the business climate will continue to evolve, as will the methods and the tools that support that evolution. Thus, the process for designing businesses will itself have to change.

Certainly, our concept of work will continue to change. Even our concept of a business will be changing. Business processes will span national as well as functional boundaries. Increasingly, global trade is viewed as essential to a company's long-term health. Thus, a company's electronics manufacturing in the Philippines will need to be linked (in terms of processes as well as systems) to its headquarters in Austin, Texas, and its sales and distribution needs throughout the world. Where one corporation begins and another ends will be less and less clear as alliances are formed and re-formed to meet changing demands. Consider the partnering between telecommunications and cable television companies (e.g., the ill-fated merger attempt of Bell Atlantic and TCI). The need for more complex products and services will cause companies to participate in an increasing number of joint ventures. Further, the lines between customer and company (as well as supplier and company) will become fuzzier as customers are provided with greater access to a company's information and its infrastructure. Who would have considered, even a *few* years ago, that a company such as Federal Express would provide customers with on-line access to its tracking system and real-time delivery status?

Our concept of organizations also will continue to evolve. As a society, we are only beginning to move out of the shadow of the industrial revolution. For example, there have been several texts on networks in organizations, and experiments with self-directed teams have provided encouraging results. However, there is still relatively little experience with these forms of organizations. In *Post-capitalist Society*, Peter Drucker foretells the continued diminution of the number of workers involved in making and moving things. He describes the workers of the future as "knowledge workers" who, organized in "task-focused teams," will use their knowledge or skill to add value to a product or a service. All that is certain is that we will be moving away from the organizational structures of Smith, Taylor, and Sloan.

Finally, our knowledge of technology and how it can make businesses more productive (and enable the creation of new businesses) is sure to change dramatically. Several technologies already exist that have the potential for such change. For example, the distribution of information will be altered forever by the mass utilization of cellular or other hand-held communications/computing devices. Given the mobility and the convenience of products recently introduced commercially, more and more businesses will turn to these technologies to change the way they share information throughout the corporation. The same may be true of information highways, which will make high-speed, long-distance communications accessible to more businesses and thus make it easier to provide the process and systems links needed between multiple locations.

All of these changes, these evolutionary steps, will impact both what we do and how we do it. Reengineering is not just for the 1990s. Rather, it recognizes that all aspects of business systems are forever changing, and that we must challenge ourselves to exchange our current environments for something different, something much better.

Appendix

The following recipes served as an example for experimental design in Chapter 11 and as a metaphor for computer instructions in Chapter 12.

STEPPER CHICKEN

1 lb. thin-sliced chicken breasts

1 egg

bread crumbs

½ tsp. marjoram

½ tsp. black pepper

fresh mozzarella (to taste)

garlic to taste

1 cup sliced mushrooms *or* 8 slices bacon

1½ cups white wine

1 tbsp. margarine or butter

Preheat oven to 375 degrees. Beat the egg and add spices. Dip each breast in the egg and coat liberally with bread crumbs. Sauté mushrooms with garlic and oil, or cook bacon. Melt butter in wine in a saucepan. Roll breaded breast with mushrooms or bacon and 1 slice of mozzarella. Cook chicken for 30 minutes, adding wine/butter occasionally. Serve with white rice and French green beans.

THE PERFECT FROZEN MARGARITA[1]

3 oz. Minute Maid Lime-Aid

3 oz. Triple Sec

6 oz. Jose Cuervo Gold Tequila

6 oz. club soda

"mucho ice"

Blend at highest speed until ice is subgranular.

---

[1]Wrangled from Jon Saxe on New Year's Eve, 1992.

Suggested Reading

PERSONAL DEVELOPMENT

Covey, Stephen R. (1989). *The Seven Habits of Highly Effective People,* Simon and Schuster. Covey, Stephen R. (1990). *Principle-centered Leadership,* Summit Books. These two books describe a comprehensive philosophy of confronting personal and professional issues. The anecdotes and techniques provided in the books will make you think about your current approach to life (career, family, etc.) and hopefully further your personal development.

Hill, Napoleon (1972). *The Think and Grow Rich Action Pack,* Hawthorn Books. This book is interesting because many of the ideas found in Covey and Robbins were expressed by Hill in the original version in 1937.

Robbins, Anthony (1991). *Awaken The Giant Within,* Summit Books. This book is the written version of many of the techniques that Robbins uses in his audiotape series on personal motivation. His ideas are tremendous building blocks for a self-mastery program.

BUSINESS IDEAS OF THE NINETIES

Drucker, Peter F. (1993). *Post-capitalist Society,* Harper Business. This insightful book offers well-thought-out ideas on society's social, political, and economic future.

Senge, Peter M. (1990). *The Fifth Discipline,* Doubleday. This was far and away our most used reference. Both of us loved this book because it is extremely thought-provoking. It is a "must read."

COGNITIVE SCIENCE

Norman, Donald A. and Stephen W. Draper (eds.) (1986). *User-Centered System Design,* Lawrence Erlbaum Associates. There are many "out-there" texts on the challenges in human-computer interaction. A former coworker, Tom Reddington, made us start to think about this topic. In many ways, it led us to really thinking about reengineering because it brought us closer to real workers. Tom recommended this book to us, and we have found many of the articles enlightening.

QUALITY MANAGEMENT

Deming, W. Edwards (1986). *Out of the Crisis,* MIT Press. Deming's style is sometimes a cross between the unpunctuated ramblings of Joseph Conrad and the incomprehensible prose of William Burroughs. However, the book is strewn with pearls of wisdom, such as the dialogues with workers and the "parable of the red beads." The chapter on special and common causes is reason enough to read this book.

Dobyns, Lloyd and Claire Crawford-Mason (1991). *Quality or Else,* Houghton Mifflin. This is a tremendous book on quality. The authors start the work by giving a brief history of the quality movement, and go on to highlight the critical differences in approach of the gurus. It is enjoyable to read.

Crosby, Philip (1979). *Quality Is Free,* McGraw-Hill. This is a seminal work on quality. We got the idea of the assessment grid from this book, and it is very enjoyable to read, as are all of Crosby's books (see the Dickens analogy in *Quality Without Tears*).

Goldratt, Eliyahu M. and Jeff Cox (1986). *The Goal,* North River Press. This excellent book, which reads like a novel, presents the fundamental concepts of statistical process control and process analysis through fictional characters and an interesting storyline. We highly recommend it for anyone interested in understanding business processes and how to improve them.

Harvey, Jerry B. (1988). *The Abilene Paradox and Other Mediations on Management,* Lexington Books. This book offers many humorous metaphors that become invaluable in explaining those business problems that have you walking away, shaking your head. You will be able to use the story of the misguided trip to Abilene at least fifty times a year.

Walton, Mary (1986). *The Deming Management Method,* Perigee Books. Mary Walton presents Deming's history, his philosophy, and evidence of his work in a clear, well-written style that is a pleasure to read.

CUSTOMER FOCUS
Whiteley, Richard C. (1991). *The Customer-Driven Company,* Addison-Wesley. Full of practical advice and helpful references, this well-written book provides useful information in an easy-to-read style.

TEAM DEVELOPMENT AND ORGANIZATION CONCEPTS
Davidow, William H. and Michael S. Malone (1992). *The Virtual Organization,* Harper Business. Businesses that successfully reengineer will be in a good position to be virtual organizations and produce virtual products. At the heart of the ideas in this book is innovation, which also is an important driver for reengineering.

Katzenbach, Jon and Douglas K. Smith (1993). *The Wisdom of Teams,* HBS Press. The authors have done some excellent research on what makes teams successful. More important, however, they succinctly present the mistakes that often are made when teams are created.

Nohria, Nitin and Robert G. Eccles (eds.) (1992). *Networks and Organizations*, HBS Press. Almost every new issue of the *Harvard Business Review* includes an article on the dynamics of informal "networks" within organizations and how they affect the overall business. Over the next decade, an understanding of networks may be as important to business success as implementing high performance teams. This book takes a step back and looks at the theoretical aspects of networks. If you are one of those people who think the network paradigm is going to explode, we recommend this book to you.

Orsburn, Jack et al. (1990). *Self-directed Work Teams*, Business One Irwin. This is a practical, hands-on guide to forming self-directed teams. It is instructive, interesting, and eminently readable.

Wellins, Richard, William Byham, and Jeanne Wilson (1991). *Empowered Teams*, Josey-Bass Inc. A history of the "team" movement, this book provides the "how-to" as well as the benefits of high performance teams.

REENGINEERING

Davenport, Thomas H. (1993). *Process Innovation*, HBS Press. This was really the first book published on reengineering. It will never sell as many copies as *Reengineering the Corporation*, but it has more detailed information on the topic. The only downside is that the style of the book (i.e., dry) makes it difficult to get through.

Davenport, Thomas and James Short. The new industrial engineering: Information technology and business process redesign, *Sloan Management Review*, Summer 1990. This article was more important to us than Hammer's initial article in *HBR*. Although Hammer gets most of the credit for pioneering reengineering, Davenport was on the scene at the same time with many of the same ideas.

Hammer, Michael, Reengineering work: Don't automate, obliterate, *Harvard Business Review,* July–August 1990. This seminal reengineering article captures what we consider to be the essence of reengineering even better than Hammer and Champy's book.

Hammer, Michael and James Champy (1993). *Reengineering the Corporation,* Harper Business. This best-selling book is perfect for getting people interested in reengineering. Although it is light on details and its treatment of technology, its clear presentation and examples make it invaluable as an introduction to reengineering.

Stalk, George et al. Competing on capabilities: The new rules of corporate strategy, *Harvard Business Review,* March–April 1992. Using different words, this article captures many of the same themes found in reengineering—shifting reward structures and top–down leadership. The article uses examples that emphasize what businesses with good leadership can accomplish, and also stresses the importance of employee capabilities.

BENCHMARKING

Camp, Robert C. (1989). *Benchmarking,* ASQC Quality Press. This easy-to-read book provides a solid overview of benchmarking and gives some detailed instructions for how to perform your own benchmarking studies. Whether you plan to do your own studies or hire consultants, this book is a good primer on the subject. Chapter 6 gives some excellent examples of benchmarking studies that highlight best-in-class characteristics.

The Verity Consulting Group (1991). *A Hands-on Guide to Competitive Benchmarking,* Verity Press. This book is not put together as well as Robert Camp's book, but it is a practical guide to benchmarking. It gives concrete suggestions on how to benchmark and includes results from a wide range of benchmarking reports.

STATISTICS

AT&T (1990). *Analyzing Business Process Data: The Looking Glass.* This book is part of the AT&T Quality Library, which is available to the general public. It is well written and contains a number of illustrative examples, making some difficult concepts accessible to readers without a background in statistics.

Cryer, Jonathan D. and Robert B. Miller (1991). *Statistics for Business: Data Analysis and Modelling,* PWS-KENT Publishing Co. This is a textbook designed for a one- or two-semester course at the college or business school level, aimed at conveying a "realistic business-world context for the use of statistical methods." It is an excellent reference text.

COMPUTER SIMULATION

Watson, Hugh J. and John H. Blackstone, Jr. (1989). *Computer Simulation,* John Wiley & Sons, Inc. Most books on simulation present more advanced theory than practical advice on how to apply the theory. This book is clearly written and provides dozens of practical examples.

EVOLUTIONARY SOFTWARE DEVELOPMENT

Arthur, Lowell Jay (1992). *Rapid, Evolutionary Development,* John Wiley & Sons, Inc. This short text gives practical reasons and guidelines for developing software using an evolutionary methodology.

Budde, Reinhard et al. (1991). *Prototyping,* Springer-Verlag. Besides contrasting traditional and evolutionary approaches, this is one of the few books that actually describe the use of specific tools to assist in prototyping. The "Technical Support of Prototyping" section (fully more than half of the book) describes everything from code generators to programming environments to organizational issues.

EXPERIMENTAL DESIGN

Box, George E. P. and Norman R. Draper (1987). *Empirical Model-Building and Response Surfaces,* John Wiley & Sons, Inc. This is a technical book, on the level of a text or reference book. It is well-written, however, and the first two chapters clearly outline the response surface methodology.

Box, George E. P., William G. Hunter, and J. Stuart Hunter (1978). *Statistics for Experimenters,* John Wiley & Sons, Inc.

DATA MODELING

Flavin, Matt (1981). *Fundamental Concepts of Information Modeling,* Yourdon Press. This is a precise definition of data modeling, intended to make modeling less of an art and more of a science. It is an excellent resource for the advanced reader interested in data analysis.

Fleming, Candace and Barbara von Halle (1989). *Handbook of Relational Database Design,* Addison-Wesley. This is a practical, easy-to-follow guide to building data models and designing databases, presented in a step-by-step fashion.

STRUCTURED ANALYSIS

Youdon, Edward (1989). *Modern Structured Analysis,* Yourdon Press. Yourdon is one of the founders of structured analysis, and in this book he provides a comprehensive look at structured methods, including advances made since structured analysis was introduced in the late 1970s. It is easy to read and would be useful for anyone interested in structured analysis.

OBJECT-ORIENTED ANALYSIS

Booch, Grady (1991). *Object-Oriented Design with Applications,* Benjamin-Cummings Publishing Company. This is one of the seminal works on object-oriented technologies. The book is meant for a technical audience, and includes a wealth of material on systems design issues.

CONCURRENT ENGINEERING

Winner, Robert I. et al. (1988). The Role of Concurrent Engineering in Weapons System Acquisition. This report was written for the Pentagon (and is available to the public because it was unclassified) "to identify critical information and factors associated with the use of concurrent engineering." Its case-study approach, including companies such as AT&T, IBM, Hewlett-Packard, and Texas Instruments, provided excellent material on what worked and what did not work in efforts to build products and services using concurrent engineering.

References

Ananda, A. L. and B. Srinivasan, eds. (1991). *Distributed Computing Systems: Concepts and Structures*, IEEE Computer Society Press.

Arthur, Lowell Jay (1992). *Rapid, Evolutionary Development*, John Wiley & Sons, Inc.

AT&T Quality Steering Committee (1991). *Reengineering Handbook*, AT&T Bell Laboratories.

Booch, Grady (1991). *Object-Oriented Design with Applications*, Benjamin-Cummings Publishing Company.

Box, George and Norman R. Draper (1987). *Empirical Model-Building and Response Surfaces*, John Wiley & Sons, Inc.

Box, George and Soren Bisgaard. The scientific context of quality improvement, *Quality Progress*, June 1987.

Brooks, Frederick P. (1975). *The Mythical Man-Month,* Addison-Wesley.

Budde, Reinhard et al. (1991). *Prototyping,* Springer-Verlag.

Camp, Robert C. (1989). *Benchmarking,* ASQC Quality Press.

Covey, Stephen R. (1990). *Principle-Centered Leadership,* Summit Books.

Covey, Stephen R. (1989). *The Seven Habits of Highly Effective People,* Simon and Schuster.

Davenport, Thomas H. (1993). *Process Innovation,* HBS Press.

Davenport, Thomas H. Need radical innovation and continuous improvement? Integrate process reengineering and TQM, *Planning Review,* May–June 1993.

Davenport, Thomas and James Short. The new industrial engineering: Information technology and business process redesign, *Sloan Management Review,* Summer 1990.

Deming, W. Edwards (1986). *Out of the Crisis,* MIT Press.

Drucker, Peter F. (1993). *Post-capitalist Society,* Harper Business.

Ernst & Young. The landmark MIT study: management in the 1990s, *Ernst & Young,* 1989.

Fleming, Candace and Barbara von Halle (1989). *Handbook of Relational Database Design,* Addison-Wesley.

Franta, W. R. (1977). *The Process View of Simulation,* Elsevier-North Holland, Inc.

General Accounting Office, Contracting for computer software development, General Accounting Office Report FGMSD-80-4, September 1979.

Gilb, T. Evolutionary design versus the waterfall model, *ACM SEN,* 10:49–61, 1985.

Goldratt, Eliyahu M. and Jeff Cox (1986). *The Goal,* North River Press.

Hammer, Michael, Reengineering work: Don't automate, obliterate, *Harvard Business Review*, July–August 1990.

Hammer, Michael and James Champy (1993). *Reengineering the Corporation*, Harper Business.

Harrington, H. James (1991). *Business Process Improvement*, McGraw-Hill, Inc.

Herzberg, Frederick. One more time: How do we motivate employees? *Harvard Business Review*, January–February 1968.

Ishikawa, Kaoru (1985). *What is Total Quality Control?*, Prentice Hall.

Ishikawa, Kaoru (1982). *Guide to Quality Control*, Quality Resources.

Katzenbach, Jon and Douglas K. Smith (1993). *The Wisdom of Teams*, HBS Press.

Loeb, Jeff (1993). Highly configured products benefit from "expert" order entry, *Manufacturing Systems*, July 1993.

Norman, Donald A. and Stephen W. Draper (eds.) (1986). *User-Centered System Design*. Lawrence Erlbaum Associates.

Peters, Tom (1987). *Thriving on Chaos*, Harper Perennial.

Senge, Peter M. (1990). *The Fifth Discipline*, Doubleday.

Shannon, R. E. (1975). *System Simulation: The Art and Science*, Prentice Hall.

Shewhart, Walter (1931). *Economic Control of Manufactured Product*, Van Nostrand.

Walton, Mary (1990). *Deming Management at Work*, Putnam.

Walton, Mary (1986). *The Deming Management Method*, Perigee Books.

Whicker, M. L. and Lee Sigelman (1991). *Computer Simulation Applications*, Sage Publications.

Whiteley, Richard C. (1991). *The Customer-Driven Company*.

Winner, Robert I. et al. (1988). The Role of Concurrent Engineering in Weapons System Acquisition (report written for the Pentagon).

Yourdon, Edward (1989). *Modern Structured Analysis,* Yourdon Press, 1989.

Zachman, J. A. A framework for information systems architecture, *IBM Systems Journal* 28(3):275–292, 1987.

Index